Privacy in Mobile and Pervasive Computing

Synthesis Lectures on Mobile and Pervasive Computing

Editor
Mahadev Satyanarayanan, *Carnegie Mellon University*

Synthesis Lectures on Mobile and Pervasive Computing is edited by Mahadev Satyanarayanan of Carnegie Mellon University. Mobile computing and pervasive computing represent major evolutionary steps in distributed systems, a line of research and development that dates back to the mid-1970s. Although many basic principles of distributed system design continue to apply, four key constraints of mobility have forced the development of specialized techniques. These include: unpredictable variation in network quality, lowered trust and robustness of mobile elements, limitations on local resources imposed by weight and size constraints, and concern for battery power consumption. Beyond mobile computing lies pervasive (or ubiquitous) computing, whose essence is the creation of environments saturated with computing and communication, yet gracefully integrated with human users. A rich collection of topics lies at the intersections of mobile and pervasive computing with many other areas of computer science.

Quality of Service in Wireless Networks Over Unlicensed Spectrum
Klara Nahrstedt
2011

The Landscape of Pervasive Computing Standards
Sumi Helal
2010

A Practical Guide to Testing Wireless Smartphone Applications
Julian Harty
2009

Location Systems: An Introduction to the Technology Behind Location Awareness
Anthony LaMarca and Eyal de Lara
2008

Replicated Data Management for Mobile Computing
Douglas B. Terry
2008

Application Design for Wearable Computing
Dan Siewiorek, Asim Smailagic, and Thad Starner
2008

Controlling Energy Demand in Mobile Computing Systems
Carla Schlatter Ellis
2007

RFID Explained: A Primer on Radio Frequency Identification Technologies
Roy Want
2006

Privacy in Mobile and Pervasive Computing
Marc Langheinrich and Florian Schaub

ISBN: 978-3-031-01358-4   paperback
ISBN: 978-3-031-02486-3   ebook
ISBN: 978-3-031-00315-8   hardcover

DOI: 10.1007/978-3-031-02486-3

A Publication in the Springer series
SYNTHESIS LECTURES ON MOBILE AND PERVASIVE COMPUTING

Lecture #13
Series Editor: Mahadev Satyanarayanan, *Carnegie Mellon University*
Series ISSN
Print 1933-9011   Electronic 1933-902X

Privacy in Mobile and Pervasive Computing

Marc Langheinrich
Università della Svizzera Italiana (USI)

Florian Schaub
University of Michigan

SYNTHESIS LECTURES ON MOBILE AND PERVASIVE COMPUTING #13

ABSTRACT

It is easy to imagine that a future populated with an ever-increasing number of mobile and pervasive devices that record our minute goings and doings will significantly expand the amount of information that will be collected, stored, processed, and shared about us by both corporations and governments. The vast majority of this data is likely to benefit us greatly—making our lives more convenient, efficient, and safer through custom-tailored and context-aware services that anticipate what we need, where we need it, and when we need it. But beneath all this convenience, efficiency, and safety lurks the risk of losing control and awareness of what is known about us in the many different contexts of our lives. Eventually, we may find ourselves in a situation where something we said or did will be misinterpreted and held against us, even if the activities were perfectly innocuous at the time. Even more concerning, privacy implications rarely manifest as an explicit, tangible harm. Instead, most privacy harms manifest as an absence of opportunity, which may go unnoticed even though it may substantially impact our lives.

In this Synthesis Lecture, we dissect and discuss the privacy implications of mobile and pervasive computing technology. For this purpose, we not only look at how mobile and pervasive computing technology affects our expectations of—and ability to enjoy—privacy, but also look at what constitutes "privacy" in the first place, and why we should care about maintaining it. We describe key characteristics of mobile and pervasive computing technology and how those characteristics lead to privacy implications. We discuss seven approaches that can help support end-user privacy in the design of mobile and pervasive computing technologies, and set forward six challenges that will need to be addressed by future research.

The prime target audience of this lecture are researchers and practitioners working in mobile and pervasive computing who want to better understand and account for the nuanced privacy implications of the technologies they are creating. Those new to either mobile and pervasive computing or privacy may also benefit from reading this book to gain an overview and deeper understanding of this highly interdisciplinary and dynamic field.

KEYWORDS

mobile computing, pervasive computing, ubiquitous computing, Internet of Things, privacy, security, privacy-enhancing technology, privacy behavior, privacy engineering

Contents

Preface

In this book, we dissect and discuss the privacy implications of mobile and pervasive computing technology. For this purpose, we not only look at how mobile and pervasive computing technology affects our expectations of—and ability to enjoy—privacy, but also look at what constitutes "privacy" in the first place, and why we should care about maintaining it. The book is structured as follows.

- **Chapter 1: Introduction.** This short chapter motivates the need for this book and outlines its contents.

- **Chapter 2: Understanding Privacy.** This chapter offers an in-depth discussion on what privacy is, i.e., what it means to "have privacy", and why we may want and need privacy. It does so by examining the concept of privacy from three perspectives: legal perspectives on privacy (also with a view towards their historic context); motivations for having (or not having) privacy; and more general conceptualizations of privacy. These perspectives support assessment of the often nuanced privacy implications of new technologies.

- **Chapter 3: Mobile and Pervasive Computing (MPC).** This chapter summarizes the key defining characteristics of mobile and pervasive computing. While mobile and pervasive computing systems feature privacy issues inherent in any computer system in general (e.g., interconnectivity), aspects such as context awareness and implicit interaction pose new privacy challenges unique to mobile and pervasive computing.

- **Chapter 4: Privacy Implications of MPC.** This chapter explores the specific privacy implications of mobile and pervasive computing in order to determine the challenges that must be addressed in order to create more privacy-friendly mobile and pervasive computing systems. It groups these around three core aspects: (1) the digitization of everyday life; (2) the ability of automatic data capture; and (3) the ability of using data to predict behavior. While none of these trends are new, mobile and pervasive computing systems exacerbate these issues greatly.

- **Chapter 5: Supporting Privacy in MPC.** This chapter discusses seven key directions and associated challenges for building privacy-friendly mobile and pervasive computing systems: (1) privacy-friendly defaults; (2) adequate privacy-risk communication; (3) privacy management assistance; (4) context-adaptive privacy mechanisms; (5) user-centric privacy controls; (6) algorithmic accountability; and (7) privacy engineering methodologies. While there is no silver bullet to remedy all privacy implications of any mobile and pervasive computing system, the presented approaches constitute an essential toolbox for building privacy into mobile and pervasive computing systems.

- **Chapter 6: Conclusions.** This chapter provides a brief outlook and stipulates key challenges for privacy that the authors see.

FOCUS AND AUDIENCE OF THIS BOOK

This book is intended as a brief introduction into the multidisciplinary area of privacy research, with a focus on its applicability to mobile and pervasive computing systems. The presented material is not meant to be comprehensive—privacy research spans a vast array of scientific disciplines and research, to which this book often only provides initial pointers. However, this book *should* provide readers with a basic understanding of the issues, complexities, and approaches involved in building privacy-aware mobile and pervasive computing systems.

The prime target audience of this lecture are researchers and practitioners working in mobile and pervasive computing who want to better understand and account for the nuanced privacy implications of the technology they are creating. Armed with the knowledge in this book, we hope they will avoid opting for simple solutions that fail to address the true complexity of the problem, or even deciding not to address privacy issues at all.

At the same time, researchers working in the areas of privacy and security in general—but without a background in mobile and pervasive systems—might want to read this lecture in order to learn about the core properties and the specific privacy challenges within the mobile and pervasive computing domains.

Last but not least, graduate and undergraduate students interested in the area should be able to gain an initial overview from this book, with enough pointers to start exploring the topic in more depth.

Marc Langheinrich and Florian Schaub
October 2018

Acknowledgments

It may come as no surprise that a project like this always takes longer than one originally antici-pates. Sometimes much longer. We are thus deeply grateful to Michael Morgan, President and CEO of Morgan & Claypool Publishers, and Mahadev "Satya" Satyanarayanan, the *Mobile and Pervasive Computing*-Series Editor, for their patience and unwavering support over the years. We also greatly benefited from the helpful feedback from both Satya and Nigel Davies, who read through countless early versions of this lecture and offered important insights on how to make this text more accessible. All of the remaining issues in this final version are fully our fault!

We also would like to thank all the staff and students at our respective universities that have supported us in our work, as well as our many collaborators near and far who help shape our research and provided us with guidance and inspiration over the years.

Marc Langheinrich and Florian Schaub
October 2018

CHAPTER 1

Introduction

In 1999, Robert Rivera slipped on some spilled yogurt in a Vons supermarket in Southern California. With a shattered kneecap as a result, Rivera sought compensation from the supermarket chain—not only to pay for his medical bills, but also to compensate for the loss of income, as he had to quit his job due to the injury. However, his effort to negotiate an out-of-court settlement fell short, according to the *LA Times* [Silverstein, 1999], when the supermarket's designated mediator produced Rivera's shopping records. Rivera was a regular Vons customer and had used their loyalty card for several years. The mediator made it clear that should this case go to court, Vons could use Rivera's shopping record to demonstrate that he regularly bought large quantities of alcohol—a fact that would surely weaken his case (who is to say that Rivera wasn't drunk when he slipped?). While Vons denied any wrongdoings, Rivera claimed that this threat prompted him to drop the case against the company.

Shopping records are a great example of the minute details that companies are interested in collecting about their customers. At first glance, it looks like a good deal: in exchange for swiping a loyalty card at the checkout,[1] consumers receive anywhere from small discounts to substantial savings on their daily grocery shopping bill. The privacy implications seem negligible. After all, the store already has a record of all items you are buying right there at checkout, so why worry about the loyalty card that helps you save money? While the difference is not obvious, the loyalty card allows for much more detailed data collection than just the payment transaction. Even though it seems as if a regular credit card not issued by the store or other cashless payment methods would be just as problematic, data flows for such cards are different: the supermarket only receives information about successful payment, but no direct identifying information about the customer; similarly, the credit card company learns that a purchase of a certain amount was made at the supermarket, but not what items were purchased. Only by also swiping a loyalty card or using a combined credit-and-loyalty card, a store is able to link a customer's identity to a particular shopping basket and thus track and analyze their shopping behavior over time.

So what is the harm? Most of us might not regularly buy "large quantities" of alcohol, so we surely would never run into the problem of Robert Rivera, where our data is used "against us". Take the case of the U.S.-American firefighter Philip Scott Lyons. A long-time customer of the Safeway supermarket chain, Lyons was arrested in August 2004 and charged with attempted arson [Schneier, 2005]. Someone had tried to set fire to Lyons' house. The fire starter found at the scene matched fire starters Lyons had previously purchased with his Safeway Club Card.

---

[1]If one uses a store-issued credit card, even that extra step disappears.

Did he start the fire himself? Luckily for Lyons, all charges against him were eventually dropped in January 2005, when another person confessed to the arson attempt. Yet for over six months, Lyons was under heavy suspicion of having set fire to his own home—a suspicion particularly damaging for a firefighter! A similar incident occurred in 2004 in Switzerland, when police found a supermarket-branded contractor's tool at the scene of a fire in the Canton of Berne. The local court forced the corresponding supermarket chain, Migros, to release the names of all 113 individuals who had bought such a tool in their stores. Eventually, all 113 suspects were removed from the investigation, as no single suspicion could be substantiated [20 Minuten].

In both the Safeway and the Migros cases, all customers who had bought the suspicious item in question (fire starters and a contractor's tool, respectively) instantly became suspects in a criminal investigation. All were ultimately acquitted of the charges against them, although particularly in the case of firefighter Lyons, the tarnished reputation that goes with such a suspicion is hard to rebuild. News stories tend to focus on suspects rather than less exciting acquittals—the fact that one's name is eventually cleared might not get the same attention as the initial suspicion. It is also often much easier to become listed in a police database as a suspect, than to have such an entry removed again after an acquittal. For example, until recently, the federal police in Switzerland would only allow the deletion of such an entry if the suspect would bring forward clear evidence of their innocence. If, however, a suspect had to be acquitted simply through lack of evidence to the contrary—as in the case of the Migros tool—the entry would remain [Rehmann, 2014].

The three cases described above are examples of privacy violations, even though none of the data disclosures (Vons' access of Robert Rivera's shopping records, or the police access of the shopping records in the US or in Switzerland) were illegal. In all three cases, data collected for one purpose ("receiving store discounts") was used for another purpose (as a perceived threat to tarnish one's reputation, or as an investigative tool to identify potential suspects). All supermarket customers in these cases thought nothing about the fact that they used their loyalty cards to record their purchases—after all, what should be so secret about buying liquor (perfectly legal if you are over 21 in the U.S.), fire starters (sold in the millions to start BBQs all around the world) or work tools? None of the customers involved had done anything wrong, yet the data recorded about them put them on the defensive until they could prove their innocence.

A lot has happened since Rivera and Lyons were "caught" in their own *data shadow*—the personal information unwittingly collected about them in companies' databases. In the 10–15 years since, technology has continued to evolve rapidly. Today, Rivera might use his Android phone to pay for all his purchases, letting not only Vons track his shopping behavior but also Google. Lyons instead might use Amazon Echo[2] to ask Alexa, Amazon's voice assistant, to order his groceries from the comfort of his home—giving police yet another shopping record to investigate. In fact, voice activation is becoming ubiquitous: many smartphones already feature

---

[2]Amazon Echo is an example of a class of wireless "smart" speakers that listen and respond to voice commands (see https://www.amazon.com/echo/); Google Home is a similar product from Google (see https://store.google.com/product/google_home).

"always-on" voice commands, which means they effectively listen in on all our conversations in order to identify a particular activation keyword.[3] Any spoken commands (or queries) are sent to a cloud server for analysis and are often stored indefinitely. Many other household devices such as TVs and game consoles[4] or home appliances and cars[5] will soon do the same.

It is easy to imagine that a future populated with an ever-increasing number of mobile and pervasive devices that record our minute goings and doings will significantly expand the amount of information that will be collected, stored, processed, and shared about us by both corporations and governments. The vast majority of this data is likely to benefit us greatly—making our lives more convenient, efficient, and safer through custom-tailored services that anticipate what we need, where we need it, and when we need it. But beneath all this convenience, efficiency, and safety lurks the risk of losing control and awareness of what is known about us in the many different contexts of our lives. Eventually, we may find ourselves in a situation like Rivera or Lyons, where something we said or did will be misinterpreted and held against us, even if the activities were perfectly innocuous at the time. Even more concerning, while in the examples we discussed privacy implications manifested as an explicit harm, more often privacy harms manifest as an absence of opportunity, which may go unnoticed even though it may substantially impact our lives.

1.1   LECTURE GOALS AND OVERVIEW

In this book we dissect and discuss the privacy implications of mobile and pervasive computing technology. For this purpose, we not only look at how mobile and pervasive computing technology affects our expectations of—and ability to enjoy—privacy, but also look at what constitutes "privacy" in the first place, and why we should care about maintaining it.

A core aspect is the question: what do we actually mean when we talk about "privacy?" Privacy is a term that is intuitively understood by everyone, but at the same time the actual meaning may differ quite substantially—among different individuals, but also for the same individual in different situations [Acquisti et al., 2015]. In the examples we discussed above, superficially, the hinging problems were the interpretation or misinterpretation of facts (Robert Rivera allegedly being an alcoholic and Philip Lyons being wrongfully accused of arson, based on their respective shopping records), but ultimately the real issue is the use of personal information for purposes not foreseen (nor authorized) originally. In those examples, privacy was thus about being "in control"—or, more accurately, the loss of control—of one's data, as well as the particular selection of facts known about oneself. However, other—often more subtle—issues exist that may rightfully be considered "privacy issues" as well. Thus, in this Synthesis Lecture we first closely

---

[3]All major smartphone platforms support such voice commands since 2015: Apple's Siri, Google Assistant, and Microsoft Cortana.

[4]Samsung TVs and the Xbox One were early devices that supported always-on voice recognition [Hern, 2015].

[5]At CES 2017, multiple companies presented voice-activated home and kitchen appliances powered by Amazon Alexa and multiple car manufactures announced integration of Amazon Alexa or Google Assistant into their new models [Laughlin, 2017].

examine the two constituents of the problem—privacy (Chapter 2) and mobile and pervasive computing technology (Chapter 3)—before discussing their intersection and illustrating the resulting challenges (Chapter 4). We finally discuss how those privacy challenges can potentially be addressed in the design of mobile and pervasive computing technologies (Chapter 5), and conclude with a summary of our main points (Chapter 6).

1.2   WHO SHOULD READ THIS

When one of the authors of this lecture was a Ph.D. student (some 15 years ago), he received a grant to visit several European research projects that worked in the context of a large EU initiative on pervasive computing—the "Disappearing Computer Initiative" [Lahlou et al., 2005]. The goal of this grant was to harness the collective experience of dozens of internationally renowned researchers that spearheaded European research in the area, in order to draft a set of "best practices" for creating future pervasive services with privacy in mind. In this respect, the visits were a failure: almost none of the half a dozen projects visited had any suggestions for building privacy-friendly pervasive systems. However, the visits surfaced an intriguing set of excuses why, as computer scientists and engineers working in the area, privacy was of no concern to them.

1. Some researchers found it best if privacy concerns (and their solutions) would be regulated socially, not technically: *"It's maybe about letting [users of pervasive technology] find their own ways of cheating."*

2. A large majority of researchers found that others where much more qualified (and required) to think about privacy: *"For [my colleague] it is more appropriate to think about [security and privacy] issues. It's not really the case in my case."*

3. Another large number of researchers thought of privacy issues simply as a problem that could (at the end) be solved trivially: *"All you need is really good firewalls."*

4. Several researchers preferred not to think about privacy at all, as this would interfere with them building interesting systems: *"I think you can't think of privacy... it's impossible, because if I do it, I have troubles with finding [a] Ubicomp future."*

With such excuses, privacy might never be incorporated into mobile and pervasive systems. If privacy is believed to be impossible, someone else's problem, trivial, or not needed, it will remain an afterthought without proper integration into the algorithms, implementations, and processes surrounding mobile and pervasive computing systems. This is likely to have substantial impact on the adoption and perception of those technologies. Furthermore, privacy laws and regulation around the world require technologists to pay attention to and mitigate privacy implications of their systems.

The prime target audience of this lecture are hence researchers and practitioners working in mobile and pervasive computing who want to better understand and account for the nuanced privacy implications of the technology they are creating, in order to avoid falling for the fallacies above. A deep understanding of potential privacy implications will help in addressing them early on in the design of new systems.

At the same time, researchers working in the areas of privacy and security in general—but without a background in mobile and pervasive systems—might want to read this lecture in order to learn about the core properties and the specific privacy challenges within the mobile and pervasive computing domains. Last but not least, graduate and undergraduate students interested in the area might want to read this synthesis lecture to get an overview and deeper understanding of the field.

CHAPTER 2

Understanding Privacy

In order to be able to appropriately address privacy issues and challenges in mobile and pervasive computing, we first need to better understand why we—as individuals and as society—might want and need privacy. What does privacy offer? How does privacy affect our lives? Why is privacy necessary? Understanding the answers to these questions naturally helps to better understand what "privacy" actually is, e.g., what it means to "be private" or to "have privacy." Only by examining the value of privacy, beyond our maybe intuitive perception of it, will we be able to understand what makes certain technology privacy invasive and how it might be designed to be privacy-friendly.

Privacy is a complex concept. Robert C. Post, Professor of Law and former dean of the Yale Law School, states that *"[p]rivacy is a value so complex, so entangled in competing and contradictory dimensions, so engorged with various and distinct meanings, that I sometimes despair whether it can be usefully addressed at all"* [Post, 2001]. In this chapter, we aim to untangle the many perspectives on and motivations for privacy. In order to better understand both the reasons for—and the nature of—privacy, we examine privacy from three perspectives. A first understanding comes from a *historical overview* of privacy, in particular from a legal perspective. Privacy law, albeit only one particular perspective on privacy, certainly is the most codified incarnation of privacy and privacy protections. Thus, it lends itself well as a starting point. Privacy law also has a rich history, with different approaches in different cultures and countries. The legal understanding of privacy has also changed substantially over the years, often because of technological advances. As we discussed in Chapter 1, technology and privacy are tightly intertwined, as technological innovations often tend to "change the playing field" in terms of making certain data practices and incursions on privacy possible that weren't possible before. Our historic overview hence also includes key moments that prompted new views on what privacy constitutes.

Our second perspective on privacy then steps back from the codification of privacy and examines arguments for and against privacy—the *motivation* for protecting or curtailing privacy. This helps us to not only understand why we may want privacy, but also what we might lose without privacy. Is privacy something valuable worth incorporating into technology?

With both the historic backdrop and privacy motivations in mind, we then present contemporary *conceptualizations* of privacy. We will see that there are many views on what privacy is, which can make it difficult to understand what someone is referring to when talking about "privacy." Precision is important when discussing privacy, in order to ensure a common understanding rather than arguing based on diverging perspectives on what privacy is or ought to be.

The discussion of different conceptualizations and understandings of privacy is meant to help us evaluate the often nuanced privacy implications of new technologies.

2.1   CODIFYING PRIVACY

There is certainly no lack of privacy definitions—in fact, this whole chapter is about defining privacy in one way or another. However, at the outset, we take a look at definitions of privacy that have received broader societal support, i.e., by virtue of being actually enshrined in law. This is not meant as legal scholarship, but rather as an overview to what are considered fundamental aspects of privacy worth protecting.

2.1.1   HISTORICAL ROOTS

Privacy is hardly a recent fad. Questions of privacy have been in the focus of society for hundreds of years. In fact, references to privacy can already be found in the Bible, e.g., in Luke 12(2–3): "What you have said in the dark will be heard in the daylight, and what you have whispered in the ear in the inner rooms will be proclaimed from the roofs" [Carroll and Prickett, 2008]. The earliest reference in common law[1] can be traced back to the English *Justices of the Peace Act* of 1361, which provided for the arrest of eavesdroppers and peeping toms [Laurant, 2003]. In 1763, William Pitt the Elder, at that time a member of the English parliament, framed in his speech on the Excise Bill the privacy of one's home as follows [Brougham, 1839]:

> The poorest man may in his cottage bid defiance to all the forces of the Crown. It may be frail—it's roof may shake—the wind may blow through it—the storm may enter—the rain may enter—but the King of England cannot enter! — all his forces dare not cross the threshold of the ruined tenement.

One of the earliest *explicit* definitions of privacy came from the later U.S. Supreme Court Justice Louis Brandeis and his colleague Samuel Warren. In 1890, the two published the essay "The Right to Privacy" [Warren and Brandeis, 1890], which created the basis for privacy tort law[2] in the U.S. legal system. They defined privacy as "the right to be let alone." The fact that this definition is so often quoted can probably be equally attributed to it being the first legal text on the subject and being easily memorizable. While it encompasses in principle all of the cases mentioned previously, such as peeping toms, eavesdroppers, and trespassers, it is still a very limited definition of privacy. Warren and Brandeis' defintion focuses on only one particular "benefit" of privacy: solitude. As we will see later in this chapter, privacy has other benefits beyond solitude.

---

[1]The *common law* is the legal system of many anglo-american countries. It is based on traditions and customs, dating back to historic England, and heavily relies on precedents. This is in contrast to "civil law" juristdictions where judgments are predominently based on codified rules.
[2]In *common law* jurisdictions, tort law governs how individuals can seek compensation for the loss or harm they experienced due to the (wrongful) actions of others.

Probably the most interesting aspect of Warren and Brandeis' work from today's perspective is what prompted them to think about the need for a legal right to privacy at the end of the 19th century:

> Recent inventions and business methods call attention to the next step which must be taken for the protection of the person, and for securing to the individual what Judge Cooley calls the right 'to be let alone.' ...Numerous mechanical devices threaten to make good the prediction that 'what is whispered in the closet shall be proclaimed from the house-tops' [Warren and Brandeis, 1890].

Figure 2.1: *The Kodak Camera*. George Eastman's "Snap Camera" made it suddenly simple to take anybody's image on a public street without their consent.

In this context, Warren and Brandeis' quote of Luke 12(2–3) (in a translation slightly different from the Bible [Carroll and Prickett, 2008]) sounds like an prescient description of the new possibilities of mobile and pervasive computing. Clearly, neither the Evangelist Luke nor Warren and Brandeis had anything like modern mobile and pervasive computing in mind. In Warren and Brandeis' case, however, it actually *was* a reference to a then novel technology—*photography*. Before 1890, getting one's picture taken usually required visiting a photographer in their studio and sitting still for a considerable amount of time, otherwise the picture would be blurred. But on October 18, 1884, George Eastmann, the founder of the Eastman Kodak Company, received U.S.-Patent #306 594 for his invention of the modern photographic film.

Instead of having to use a large tripod-mounted camera with heavy glass plates in the studio, everybody could now take Kodak's "Snap Camera" (see Figure 2.1) out to the streets and take a snapshot of just about anybody—without their consent. It was this rise of unsolicited pictures, which more and more often found their way into the pages of the (at the same time rapidly expanding) tabloid newspapers, that prompted Warren and Brandeis to paint this dark picture of a world without privacy.

Today's developments of smartphones, wearable devices, smart labels, memory amplifiers, and IoT-enabled smart "things" seem to mirror the sudden technology shifts experienced by Warren and Brandeis, opening up new forms of social interactions that change the way we experienced our privacy in the past. However, Warren and Brandeis' "right to be let alone" looks hardly practical today: with the multitude of interactions in today's world, we find ourselves constantly in need of dealing with people (or better: services) that do not know us in person, hence require some form of personal information from us in order to judge whether such an interaction would be beneficial. From opening bank accounts, applying for credit, obtaining a personal yearly pass for trains or public transportation, or buying goods online—we constantly have to "connect" with others (i.e., give out our personal information) in order to participate in today's life. Even when we are not explicitly providing information about ourselves we constantly leave digital traces. Such traces range from what websites we visit or what news articles we read, to surveillance and traffic cameras recording our whereabouts, to our smartphones revealing our location to mobile carriers, app developers and advertisers. Preserving our privacy through isolation is just not as much of an option anymore as it was over a 100 years ago.

Privacy as a Right

Warren and Brandeis' work put privacy on the legal map, yet it took another half century before privacy made further legal inroads. After the end of the Second World War, in which Nazi Germany had used detailed citizen records to identify unwanted subjects of all kinds [Flaherty, 1989], privacy became a key human right across a number of international treaties—the most prominent being the Universal Declaration of Human Rights, adopted by the United Nations in 1948, which states in its Article 12 that [United Nations, 1948]:

> No one shall be subjected to arbitrary interference with his privacy, family, home or correspondence, nor to attacks upon his honor and reputation. Everyone has the right to the protection of the law against such interference or attacks.

Similar protections can be found in Article 8 of the Council of Europe's Convention of 1950 [Council of Europe, 1950], and again in 2000 with the European Union's Charter of Fundamental Rights [European Parliament, 2000], which for the first time in the European Union's history sets out in a single text the whole range of civil, political, economic, and social rights of European citizens and all persons living in the European Union [Solove and Rotenberg, 2003]. Article 8 of the Charter, concerning the Protection of Personal Data, states the following [European Parliament, 2000].

1. Everyone has the right to the protection of personal data concerning him or her.

2. Such data must be processed fairly for specified purposes and on the basis of the consent of the person concerned or some other legitimate basis laid down by law. Everyone has the right of access to data which has been collected concerning him or her, and the right to have it rectified.

3. Compliance with these rules shall be subject to control by an independent authority.

The rise of the Internet and the World Wide Web in the early 1990s had prompted many to proclaim the demise of national legal frameworks, as their enforcement in a borderless cyberspace seemed difficult at least.[3] However, the opposite effect could be observed: at the beginning of the 21st century, many national privacy laws have not only been adjusted to the technical realities of the Internet, but also received a substantial international harmonization facilitating cross-border enforcement.

Today, more than 100 years after Warren and Brandeis laid the foundation for modern data protection laws, two distinctive principles for legal privacy protection have emerged: the European approach of favoring comprehensive, all-encompassing data protection legislation that governs both the private and the public sector, and the sectoral approach popular in the United States that favors sector-by-sector regulation in response to industry-specific needs and concerns in conjunction with voluntary industry self-regulation. In both approaches, however, privacy protection is broadly modeled around what is known as "Fair Information Practice Principles."

The Fair Information Practice Principles

If one would want to put a date to it, modern privacy legislation was probably born in the late 1960s and early 1970s, when governments first began to systematically make use of computers in administration. Alan Westin's book *Privacy and Freedom* published in 1967 [Westin, 1967] had a significant impact on how policymakers in the next decades would address privacy. Clarke [2000] reports how a 1970 German translation of Westin's book significantly influenced the world's first privacy law, the "Datenschutzgesetz" (data protection law) of the West German state Hesse. In the U.S., a Westin-inspired 1973 report of the *United States Department for Health Education and Welfare* (HEW) set forth a code of *Fair Information Practice* (FIP), which has become a cornerstone of U.S. privacy law [Privacy Rights Clearinghouse, 2004], and has become equally popular worldwide. The five principles are as follows [HEW Advisory Committee, 1973].

1. There must be no personal data record keeping systems whose very existence is secret.

---

[3]In his 1996 "Declaration of Independence of Cyberspace," John Barlow, co-founder of the Electronic Frontier Foundation (EFF), declared "Governments of the Industrial World, you weary giants of flesh and steel, I come from Cyberspace, the new home of Mind. On behalf of the future, I ask you of the past to leave us alone. You are not welcome among us. You have no sovereignty where we gather" [Barlow, 1996].

2. There must be a way for an individual to find out what information about him is in a record and how it is used.

3. There must be a way for an individual to prevent information about him that was obtained for one purpose from being used or made available for other purposes without his consent.

4. There must be a way for an individual to correct or amend a record of identifiable information about him.

5. Any organization creating, maintaining, using, or disseminating records of identifiable personal data must assure the reliability of the data for their intended use and must take precautions to prevent misuse of the data.

In the early 1980s, the Organization for Economic Cooperation and Development (OECD) took up those principles and issued "The OECD Guidelines on the Protection of Privacy and Transborder Flows of Personal Data" [OECD, 1980], which expanded them into eight practical measures aimed at harmonizing the processing of personal data in its member countries. By setting out core principles, the organization hoped to *"obviate unnecessary restrictions to transborder data flows, both on and off line."* The eight principles are as follows [OECD, 2013].[4]

1. *Collection Limitation Principle.* There should be limits to the collection of personal data and any such data should be obtained by lawful and fair means and, where appropriate, with the knowledge or consent of the data subject.

2. *Data Quality Principle.* Personal data should be relevant to the purposes for which they are to be used, and, to the extent necessary for those purposes, should be accurate, complete and kept up-to-date.

3. *Purpose Specification Principle.* The purposes for which personal data are collected should be specified not later than at the time of data collection and the subsequent use limited to the fulfillment of those purposes or such others as are not incompatible with those purposes and as are specified on each occasion of change of purpose.

4. *Use Limitation Principle.* Personal data should not be disclosed, made available or otherwise used for purposes other than those specified in accordance with the Purpose Specification principle except:

    (a) with the consent of the data subject; or

---

[4]In 2013, the OECD published a revision of its privacy guidelines [OECD, 2013]. The OECD privacy principles remain unchanged in the 2013 version, except for a gender neutral reformulation of the individual participation principle, which we provide here. The revisions primarily updated the OECD's recommendations regarding the principles' implementation with a focus on the practical implementation of privacy protection through risk management and the need to address the global dimension of privacy through improved interoperability.

(b) by the authority of law.

5. *Security Safeguards Principle.* Personal data should be protected by reasonable security safe-guards against such risks as loss or unauthorised access, destruction, use, modification or disclosure of data.

6. *Openness Principle.* There should be a general policy of openness about developments, prac-tices and policies with respect to personal data. Means should be readily available of es-tablishing the existence and nature of personal data, and the main purposes of their use, as well as the identity about usual residence of the data controller.[5]

7. *Individual Participation Principle.* Individuals should have the right:

   (a) to obtain from a data controller, or otherwise, confirmation of whether or not the data controller has data relating to them;

   (b) to have communicated to them, data relating to them

      i. within a reasonable time;

      ii. at a charge, if any, that is not excessive;

      iii. in a reasonable manner; and

      iv. in a form that is readily intelligible to them;

   (c) to be given reasons if a request made under subparagraphs (a) and (b) is denied, and to be able to challenge such denial; and

   (d) to challenge data relating to them and, if the challenge is successful, to have the data erased; rectified, completed or amended.

8. *Accountability Principle.* A data controller should be accountable for complying with mea-sures which give effect to the principles stated above.

Even though the OECD principles, just as the HEW guidelines before them, carried no legal obligation, they nevertheless constituted an important international consensus that sub-stantially influenced national privacy legislation in many countries in the years to come [Solove and Rotenberg, 2003]. In what Michael Kirby, former Justice of the High Court in Australia, has called the "decade of privacy" [Clarke, 2006], many European countries (and the U.S.) fol-lowed the German state Hesse in passing comprehensive data protection laws—the first national privacy law was passed in Sweden in 1973, followed by the U.S. (Privacy Act of 1974, regulating processing of personal information by federal agencies), Germany (1977), and France (1978).

---

[5]Privacy law typically considers three principal roles: data subjects, data controllers, and data processors. Data *subjects* are natural persons whose personal data (i.e., "any information relating to an identified or identifiable natural person") is collected and processed; a data controller is "the natural or legal person, public authority, agency or other body which, alone or jointly with others, determines the purposes and means of the processing of personal data;" a data *processors* is "a natural or legal person, public authority, agency or other body which processes personal data on behalf of the controller" [European Parliament and Council, 2016].

The FIPs, while an important landmark in privacy protection, are, however, not without their flaws. Clarke [2000] calls them a "movement that has been used by corporations and governments since the late 1960s to avoid meaningful regulation." Instead of taking a holistic view on privacy, Clark finds the FIPs too narrowly focused on "data protection," only targeting the "facilitation of the business of government and private enterprise" rather than the human rights needs that should be the real goal of privacy protection: "the principles are oriented toward the protection of data about people, rather than the protection of people themselves" [Clarke, 2006]. More concrete omissions of the FIPs are the complete lack of data deletion or anonymization requirements (i.e., after the data served its purpose), or the absence of clear limits on what could be collected and in what quantities (the FIPs only require that the data collected is "necessary"). Similarly, Cate [2006] notes that, in their translation into national laws, the broad and aspirational fair information practice principles have often been reduced to narrow legalistic concepts, such as notice, choice, access, security, and enforcement. These narrow interpretations of the FIPs focus on procedural aspects of data protection rather than the larger goal of protecting privacy for the benefit of individuals and society.

2.1.2   PRIVACY LAW AND REGULATIONS

Many countries have regulated privacy protections through national laws—often with reference to or based on the fair information practice principles. We provide an overview of those laws with a specific emphasis on the U.S. and Europe, due to their prominent roles in developing and shaping privacy law and their differing approaches for regulating privacy.

Privacy Law and Regulations in the United States
The U.S. Constitution does not lay out an explicit constitutional right to privacy. However, in a landmark case, *Griswold vs. Connecticut* 1965,[6] the U.S. Supreme Court recognized a constitutional right to privacy, emanating from the First, Third, Fourth, Fifth, and Ninth Amendments of the U.S. Constitution.[7] The First Amendment guarantees freedom of worship, speech, press, assembly and petition. Privacy under First Amendment protection usually refers to being unencumbered by the government with respect to one's views (e.g., being able to speak anonymously or keeping one's associations private). The Third Amendment provides that troops may not be quartered (i.e., allowed to reside) in private homes without the owner's consent (an obvious relationship to the privacy of the home). The Ninth Amendment declares that the listing of

---

[6] *Griswold vs. Connecticut* involved the directors of the Planned Parenthood League of Connecticut, a nonprofit agency which disseminated birth control information, who challenged a Connecticut law criminalizing contraceptives and counseling about contraceptives to married couples. The Court held that the law was unconstitutional, and specifically described two interests for protecting privacy: (1) "the individual interest in avoiding disclosure of personal matters" and (2) "the interest in independence in making certain kinds of important decisions" [Solove and Rotenberg, 2003].

[7] The first ten amendments to the U.S. Constitution are collectively known as the "Bill of Rights." They were added as a result of objections to the original Constitution of 1787 during state ratification debates. Congress approved these amendments as a block of twelve in September 1789, and the legislatures of enough states had ratified 10 of those 12 by December 1791 [Wikipedia].

individual rights is not meant to be comprehensive, i.e., that the people have other rights not specifically mentioned in the Constitution [National Archives]. The right to privacy is primarily anchored in the Fourth and Fifth Amendments [Solove and Rotenberg, 2003].

- *Fourth Amendment:* The right of the people to be secure in their persons, houses, papers, and effects, against unreasonable searches and seizures, shall not be violated, and no Warrants shall issue, but upon probable cause, supported by Oath or affirmation, and particularly describing the place to be searched, and the persons or things to be seized.

- *Fifth Amendment:* No person shall be […] compelled in any criminal case to be a witness against himself, nor be deprived of life, liberty, or property, without due process of law; nor shall private property be taken for public use, without just compensation.

In addition, the Fourteenth Amendment's due process clause has been interpreted to provide a substantive due process right to privacy.[8]

- *Fourteenth Ammendment:* No state shall make or enforce any law which shall abridge the privileges or immunities of citizens of the United States; nor shall any state deprive any person of life, liberty, or property, without due process of law; nor deny to any person within its jurisdiction the equal protection of the laws.

While the U.S. Constitution recognizes an individual right to privacy, the constitution only describes the rights of citizens in relationship to their government, not to other citizens or companies[9] [Cate, 1997]. So far, no comprehensive legal privacy framework exists in the United States that equally applies to both governmental and private data processors. Instead, federal privacy law and regulation follows a sectoral approach, addressing specific privacy issues that arise in certain public transactions or industry sectors [Solove and Schwartz, 2015].

Privacy with respect to the government is regulated by the *Privacy Act of 1974*, which only applies to data processing at the federal level [Gormley, 1992]. The Privacy Act roughly follows the Fair Information Principles set forth in the HEW report (mentioned earlier in this section), requiring government agencies to be transparent about their data collections and to support access rights. It also restricts what information different government agencies can share about an individual and allows citizens to sue the government for violating these provisions. Additional laws regulate data protection in other interactions with the government, such as the *Driver's Privacy Protection Act (DPPA)* of 1994, which restricts states in disclosing or selling personal information from motor vehicle records, or the Electronic Communications Privacy Act (ECPA) of 1986, which extended wiretapping protections to electronic communication.

---

[8]That the Fourteenth Amendment provides a due process right to privacy was first recognized in concurring opinions of two Supreme Court Justices in *Griswold v. Connecticut*. It was also recognized in *Roe v. Wade* 1973, which invoked the right to privacy to protect a woman's right to an abortion, and in *Lawrence v. Texas* 2003, which invoked the right to privacy regarding the sexual practices of same-sex couples.

[9]An exception is the 13th Amendment, which prohibits slavery and thus also applies to private persons.

Privacy regulation in the private sector is largely based on self-regulation, i.e., industry associations voluntarily enact self-regulations for their sector to respect the privacy of their customers. In addition, federal or state privacy laws are passed for specific industry sectors in which privacy problems emerge. For instance, the Family Educational Rights and Privacy Act (FERPA) of 1974 regulates student privacy in schools and universities; and the Children's Online Privacy Protection Act (COPPA) of 1998 restricts information collection and use by websites and online services for children under age 13.

The Health Insurance Portability and Accountability Act (HIPAA) of 1996 gives the Department of Health and Human Services rule making authority regarding the privacy of medical records. The HIPAA Privacy Rule requires privacy notices to patients, patient authorization for data processing and sharing, limits data processing to what is necessary for healthcare, gives patients data access rights, and prescribes physical and technical safeguards for health records. Commonly, federal privacy laws are amended over time to account for evolving privacy issues. For instance the Genetic Information Nondiscrimination Act (GINA) of 2008 limits the use of genetic information in health insurance and employment decisions.

Privacy in the financial industry is regulated by multiple laws. The Fair Credit Reporting Act (FCRA) of 1970 governs how credit reporting agencies can use consumer information. It has been most recently amended by the Economic Growth, Regulatory Relief, and Consumer Protection Act of 2018, which, as a reaction to the 2017 Equifax Data Breach, gave consumers the right to free credit freezes to limit access to their credit reports and thus reduce the risk of identity theft. The Gramm-Leach-Bliley Act (GLBA) of 1999 requires that financial institutions store financial information in a secure manner, provide customers with a privacy notice annually and gives consumers the right to opt-out or limit sharing of personal information with third parties.

The Telephone Consumer Protection Act (TCPA) of 1991 provides remedies from repeat telephone calls by telemarketers and created the national Do Not Call registry.[10] The Controlling the Assault of Non-Solicited Pornography And Marketing (CAN-SPAM) Act of 2003 created penalties for the transmission of unsolicited email and requires that email newsletters and marketing emails must contain an unsubscribe link. The Video Privacy Protection Act (VPPA) of 1988 protects the privacy of video rental records.

Those federal privacy laws are further complemented by state laws. For instance, many states have passed RFID-specific legislation that prohibits unauthorized reading of RFID-enabled cards and other devices (e.g., the state of Washington's Business Regulation Chapter 19.300 [Washington State Legislature, 2009]). The state of Delaware enacted four privacy laws in 2015, namely the Online and Personal Privacy Protection Act (DOPPA), the Student Data Privacy Protection Act (SDPPA), the Victim Online Privacy Act (VOPA), and the Employee/Applicant Protection for Social Media Act (ESMA).

---

[10]National Do Not Call Registry: https://www.donotcall.gov.

One of the more well-known state privacy laws is California's Online Privacy Protection Act (CalOPPA) of 2004, which poses transparency requirements, including the posting of a privacy policy, for any website or online service that collects and maintains personally identifiable information from a consumer residing in California. Because California is the most populous U.S. state with a large consumer market and due to the difficulty of reliably determining an online user's place of residence, CalOPPA, despite being a state law, affected almost all US websites as well as international websites. In 2018, California became the first US state to enact a comprehensive (i.e., non-sectoral) privacy law. The California Consumer Privacy Act of 2018, which will go into effect in 2020, requires improved privacy notices, a conspicuous opt-out button regarding the selling of consumer information, and grants consumers rights to data access, deletion and portability.

Due to the fractured nature of privacy legislation, privacy enforcement authority is also divided among different entities, including the Department of Health and Human services (for HIPAA), the Department of Education (for FERPA), State Attorneys General (for respective state laws), and the Federal Trade Commission (FTC). The FTC, as the U.S. consumer protection agency, has a prominent privacy enforcement role [Solove and Hartzog, 2014], including the investigation of deceptive and unfair trade practices with respect to privacy, as well as statutory enforcement (e.g., for COPPA). The FTC further has enforcement power with respect to Privacy Shield, the U.S.–European agreement for cross-border transfer. Due to its consumer protection charge, the FTC can also bring privacy-related enforcement actions against companies in industries without a sectoral privacy law [Solove and Hartzog, 2014], such as mobile apps, online advertising, or smart TVs. In addition to monetary penalties, FTC consent decrees typically require companies to submit to independent audits for 20 years and to establish a comprehensive internal security or privacy program. The FTC's enforcement creates pressure for industries to adhere to their self-regulatory privacy promises and practices.

In addition to federal and state laws, *civil* privacy lawsuits (i.e., between persons or corporations) are possible. Prosser [1960] documented four distinct privacy *torts* common in US law,[11] i.e., ways for an individual who felt their privacy has been violated to sue the violator for damages:

- *intrusion* upon seclusion or solitude, or into private affairs;

- *public disclosure* of embarrassing private facts;

- *adverse publicity* which places a person in a false light in the public eye; and

- *appropriation* of name of likeness.

---

[11]A *tort* is a civil wrong for which the law provides remedy. The "law of torts" is part of the *common law*, which is the legal system of many Anglo-American countries, such as the UK or the U.S. In contrast to *civil law* practiced in most European countries (which is derived from Roman law, and has the form of statutes and codes written and enacted by emperors, kings, and, today, by national legislatures), common law is based on traditions, customs, and precedents dating back to historical England.

In summary, privacy is protected in the U.S. by a mix of sector-specific federal and state laws, with self-regulatory approaches and enforcement by the FTC in otherwise unregulated sectors. An advantage of this sectoral approach is that resulting privacy laws are often specific to the privacy issues, needs, and requirements in a given sector, a downside is that laws are often surpassed by the advancement of technology, thus requiring periodical amendments.

Privacy Law and Regulation in the European Union

On the other side of the Atlantic, a more civil-libertarian perspective on personal data protection prevails. Individual European states began harmonizing their national privacy laws as early as the mid-1970s. In 1973 and 1974, the European Council[12] passed resolutions (73)22 and (74)29, containing guidelines for national legislation concerning private and public databases, respectively [Council of Europe, 1973, 1974]. In 1985, the "Convention for the Protection of Individuals with regard to Automatic Processing of Personal Data" (108/81) went into effect, providing a normative framework for national privacy protection laws of its member states [Council of Europe, 1981]. Convention 108/81 is open to any country to sign (i.e., not only CoE members), and has since seen countries like Uruguay, Mauritus, Mexico, or Senegal join.[13] While the convention offered a first step toward an international privacy regime, its effect on national laws remained relatively limited [Mayer-Schönberger, 1998].

It was the 1995 Data Protection Directive 95/46/EC [European Parliament and Council, 1995] (in the following simply called "the Directive") that achieved what Convention 108/81 set out to do, namely a lasting harmonization of the various European data protection laws and providing an effective international tool for privacy protection even beyond European borders.

The Directive had two important aspects that advanced its international applicability. On the one hand, it required all EU member states[14] to enact national law that provided at least the same level of protection as the Directive stipulated. This European harmonization allowed for a free flow of information among all its member states, as personal data enjoyed the same minimum level of protection set forth by the Directive in any EU country.

On the other hand, the Directive's Article 25 explicitly prohibited the transfer of personal data into "unsafe third countries," i.e., countries with data protection laws that would not offer an adequate level of protection as required by the Directive. After European officials made it clear that they intended to pursue legal action against the European branch offices of corporations that would transfer personal data of EU residents to their corresponding headquarters in such unsafe third countries, a large number of non-European countries around the world

---

[12]The European Council was founded in 1949 in order to harmonize legal and social practices across Europe. It groups together 47 countries—the 28 EU member states and additional mostly central and eastern European countries. Since 1989, its main job has become assisting the post-communist democracies in central and eastern Europe in carrying out political, legal, and economic reform.

[13]As of August 2018, all 47 CoE member states have ratified the convention, while six non-member states have done so; see `conventions.coe.int/Treaty/Commun/ChercheSig.asp?NT=108&CM=8&DF=8/18/04&CL=ENG` for latest figures.

[14]The directive actually applies to the so-called "European Economic Area" (EEA), which not only includes the EU-member states but also Norway, Iceland, and Liechtenstein.

began to adjust their privacy laws in order to become a "safe"country with regards to the Directive, and thus become part of the European Internal Information Market. Eventually, a dozen countries were considered "safe" third-countries with respect to personal data transfers: Andorra, Argentina, Canada, Switzerland, Faeroe Islands, the British Channel Islands (Guernsey, Jersey, Isle of Man), Israel, New Zealand, the U.S.,[15] and Uruguay.

However, despite its significant impact, the 1995 Directive was woefully ignorant of the rapid technological developments of the late 1990s and early 2000s. It was created before the Web took off, before smartphones appeared, before Facebook and Twitter and Google were founded. It is not surprising then that many criticized it for being unable to cope with those realities [De Hert and Papakonstantinou, 2012]. While the Directive was specifically written to be "technology neutral," it also meant that it was unclear how it would apply to many concrete technical developments, such as location tracking, Web cookies, online profiling, or cloud computing. In order to bring the European privacy framework more in line with the realities of mobile and pervasive computing, as well as to create a single data protection law that applies in all EU member states, an updated framework was announced in 2012 and finally enacted in early 2016—the *General Data Protection Regulation* (GDPR). The GDPR then went into effect on May 25, 2018. Its main improvements over the 1995 Directive can be summarized as follows [De Hert and Papakonstantinou, 2012, 2016].

1. *Expanded Coverage:* As per its Article 3, the GDPR now also applies to companies outside of the EU who offer goods or services to customers in the EU ("marketplace rule")—the 1995 Directive only applied to EU-based companies (though it attempted to limit data flows to non EU-based companies).

2. *Mandatory Data Protection Officers (DPO):* Article 37 requires companies whose "core activities... require regular and systematic monitoring of data subjects on a large scale" to designate a DPO as part of their accountability program, who will be the main contact for overseeing legal compliance.

3. *Privacy by Design:* Article 25 requires that all data collection and processing must now follow a "data minimization" approach (i.e., collect only as much data as absolutely necessary), that privacy is provided by default, and that entities use detailed impact assessment procedures to evaluate the safety of its data processing.

4. *Consent:* Article 7 stipulates that those who collect personal data must demonstrate that it was collected with the consent of the data subject, and if the consent was "freely given." For example, if a particular piece of data is not necessary for a service, but if the service is withheld from a customer otherwise, would not qualify as "freely given consent."

---

[15]Data transfer between Europe and the U.S. has been regulated by a separate agreement called the Safe Harbor Agreement, which was later replaced by the EU-US Privacy Shield (see `https://www.privacyshield.gov/`) [Weiss and Archick, 2016].

5. *Data Breach Notifications:* Article 33 requires those who store personal data to notify national data protection authorities if they are aware of a "break-in" that might have resulted in personal data being stolen. Article 34 extends this to also notify data subjects if the breach "is likely to result in a high risk to the rights and freedoms of natural persons."

6. *New Subject Rights:* Articles 15–18 give those whose data is collected more explicit rights, such as the right to object to certain uses of their data, the right to obtain a copy of the personal data undergoing processing, or the right to have personal data being deleted ("the right to be forgotten").

How these changes will affect privacy protection in Europe and beyond will become clearer over the coming years. When the GDPR finally came into effect in May 2018, its most visible effect was a deluge of email messages that asked people to confirm that they still wanted to be on a mailing list (i.e., giving "unambiguous" consent, as per Article 4) [Hern, 2018, Jones, 2018], as well as a pronounced media backlash questioning both the benefits of the regulation [Lobo, 2018] as well as its (seemingly extraordinarily high) costs [Kottasová, 2018]. Many of the new principles in the GDPR sound simple, but can be challenging to implement in practice (e.g., privacy by design, the right to erasure). We will discuss some of these challenges in Chapter 6. Also, the above-mentioned Council of Europe "Convention for the Protection of Individuals with regard to Automatic Processing of Personal Data" (108/81) [Council of Europe, 1981] has recently been updated [Council of Europe, 2018] and is now being touted as a first step for non-EU countries to receive the coveted status of a "safe third country" (adequacy assessment) [European Commission, 2017] with respect to the new GDPR [Greenleaf, 2018].

Privacy Law and Regulation in Other Countries
Beyond the U.S. and Europe, many countries have adopted data protection or privacy laws [Greenleaf, 2017, Swire and Ahmad, 2012]. An increasing number of countries have been adopting comprehensive data protection laws, which not just follow the Europan model, but are often based on EU Directive 95/46/EC or the GDPR. For instance, the data protection laws of Switzerland, Russia, and Turkey are similar to the EU Directive. Mexico's 2010 Federal Law on the Protection of Personal Data Held by Private Entities also follows a comprehensive approach similar to the EU Directive, in particular with respect to data subjects' rights, obligations of data controllers and processors, and international data transfer requirements. The Mexican law further incorporates the Habeas Data concept common in Latin American legal regimes [Swire and Ahmad, 2012]. Habeas Data refers to the constitutional right that citizens "may have the data" that is stored about them, i.e., they have the right to pose habeas data requests to entities to learn whether and what information is stored about them and request correction. The Mexican law requires data controllers to designate a contact for such requests and process them in a timely manner. The GDPR's data portability right (Art. 20, GDPR) provides a similar right for data subjects and obligations for data controllers. In 2018, Brazil adopted the General Data

Privacy Law (LGPD), which goes into effect in 2020. The LGPD closely mirrors the GDPR in its key provisions.

Canada also employs a comprehensive data protection approach. PIPEDA, the Personal Information Protection and Electronic Documents Act, regulates data protection for the private sector in Canada. A key difference between the GDPR and PIPEDA is that under PIPEDA individual informed consent is the only basis for lawful data collection, processing, and sharing, with limited exceptions [Banks, 2017].

Australia employs a co-regulatory model. Australia's Federal Privacy Act defines National Privacy Principles for government agencies and the private sector. Industries then define self-regulatory codes that reflect the National Privacy Principles, with oversight by the Australian National Privacy Commissioner.

The Privacy Framework of the Asia-Pacific Economic Cooperation (APEC) aims to promote interoperability of privacy regimes across the 21 APEC countries. In contrast to Europe's GDPR, the APEC Privacy Framework [APEC, 2017] is not a law but rather defines nine privacy principles, based on the OECD privacy guidelines, APEC countries can choose to subscribe to. The Framework further defines Cross-Border Privacy Rules (CBPR) as a code of conduct to enable cross-border data transfers among countries committing to the CBPR. The CBPR requires a local accountability agent (i.e., a governmental institution) that certifies organization's CBPR compliance. As of 2018, six APEC countries are participating in CBPR, namely the U.S., Japan, Mexico, Canada, South Korea, and Singapore. In addition to the CBPR, the APEC Cross-border Privacy Enforcement Agreement (CPEA) facilitates cooperation and information sharing among APEC countries' privacy enforcement authorities.

2.2   MOTIVATING PRIVACY

When the UK government in 1994 tried to rally support for its plans to significantly expand CCTV surveillance in Britain, it coined the slogan *"If you've got nothing to hide, you've got nothing to fear"* [Rosen, 2001]—a slogan that has been a staple in counter-privacy arguments ever since. What is so bad of having less privacy in today's day and age, unless you are a terrorist, criminal, or scoundrel? Surely, people in Britain, with its over 6 million surveillance cameras (one for every 11 people) [Barrett, 2013] seem to be no worse off than, say, their fellow European neighbors in France or Germany, which both have nowhere near that many cameras.[16] Would those who maintain an active Facebook page say they are worse off than those who only use email, text messages, or, say, written letters, to communicate with friends and family? Why not let Google monitor all Web searches and emails sent and received, so that it can provide better search results, a cleaner inbox, and more relevant targeted advertising, rather than the random spam that usually makes it into one's inbox? Who would not want police and other national security

---

[16]Note that surveillance cameras seem to have no significant effect on violent crime—one of the main reasons for having all those cameras in the first place. According to a 2013 report from the UK College of Policing [National Police Library – College of Computing, 2013], CCTV leads only to "a small reduction in crime," mostly in the area of vehicle theft, but that it has "no impact on the levels of violent crimes."

institutions have access to our call records and search history in order to prevent terrorists and child molesters from planning and conducting their heinous crimes?

One might assume that making the case for privacy should be easy. Privacy is one of the leading consumer concerns on the Internet, dominating survey responses for more than 20 years now (e.g., Westin's privacy surveys between 1990 and 2003 [Kumaraguru and Cranor, 2005], the 1999 IBM Multi-National Consumer Privacy Survey [IBM Global Services, 1999], or recent consumer reports from KPMG [2016] or International Data Corporation (IDC) [2017]). Everybody seems to *want* privacy. However, when separating preferences from actual behavior [Berendt et al., 2005, Spiekermann et al., 2001], most people in their everyday life seem to care much less about privacy than surveys indicate—something often called the "privacy paradox" [Norberg et al., 2007]. Facebook, with its long history of privacy-related issues [Parakilas, 2017], is still growing significantly every year, boasting over 2.23 billion "active monthly users"[17] at the end of June 2018 [Facebook, Inc., 2018]. Back in 2013, with only about half that many active users (1.2 billion) [Facebook, Inc., 2018], Facebook users already shared almost 3.3 million pieces of content (images, posts, links) per minute [Facebook, Inc., 2013]. Within the same 60 s, Google serves an estimated 3.6 million search queries [James, 2017], each feeding into the profile of one of its over 1+ billion unique users[18] in order to better integrate targeted advertising into their search results, Gmail inboxes, and YouTube videos. Of course, more privacy-friendly alternatives exist and they do see increasing users. For example, a service like the anonymous search engine DuckDuckGo saw its traffic double within days[19] after Edward Snowden revealed the extent to which many Internet companies, including Google, were sharing data with the U.S. government. However, DuckDuckGo's share of overall searches remains minuscule. Even though its share had been on the rise ever since the Snowden leaks of June 2013, its current[20] 11 million queries a day (roughly seven times its pre-Snowden traffic) are barely more than 0.3%[21] of Google's query traffic.

Why are not more people using a privacy-friendly search engine like DuckDuckGo? Does this mean people do not care about privacy? Several reasons come to mind. First, not many people may have heard about DuckDuckGo. Second, "traditional" search engines might simply provide superior value over their privacy-friendly competitors. Or maybe people simply think that they do. Given that the apparent cost of the services is the same (no direct charge to the consumer), the fact that one offers more relevant results than the other may be enough to make people not want to switch. Third, and maybe most important: indirect costs like a loss of privacy are notoriously hard to assess [Solove, 2013]. What could possibly happen if Yahoo, Microsoft, or Google know what one is searching? What is so bad about posting holiday pictures on Facebook

---

[17]An active user is someone who has logged in at least once in the last 30 days.
[18]As announced at Google's 2017 I/O developer conference [Popper, 2017].
[19]See https://duckduckgo.com/traffic.html.
[20]As of January 2017, see https://duckduckgo.com/traffic.html.
[21]According to Internetlivestats.com [2018], Google serves over 3.5 billion queries a day. Google does not publicly disclose the number of queries they serve.

or Instagram? Why would chatting through Signal[22] be any better than through WhatsApp?[23] Consider the following cases.

- In 2009, U.S. Army veteran turned stand-up comedian Joe Lipari had a bad customer experience in his local Apple store [Glass, 2010]. Maybe unwisely, Joe went home and took out his anger via a Facebook posting that quoted a line from the movie he started watching—*Fight Club* (based on the 1996 book by Palahniuk [1996]): "And this button-down, Oxford-cloth psycho might just snap, and then stalk from office to office with an Armalite AR-10 carbine gas-powered semi-automatic weapon, pumping round after round into colleagues and co-workers." Lipari posted the slightly edited variant: *Joe Lipari might walk into an Apple store on Fifth Avenue with an Armalite AR-10 carbine gas-powered semi-automatic weapon and pump round after round into one of those smug, fruity little concierges.*" An hour later, a full SWAT team arrived, apparently alerted by one of Joe's Facebook contacts who had seen the posting and contacted homeland security. After a thorough search of his place and a three-hour interrogation downtown, Joe assumed that his explanation of this being simply a bad movie quote had clarified the misunderstanding. Yet four months later, Joe Lipari was charged with two "Class D" felonies—"PL490.20: Making a terroristic threat" [The State of New York, 2018b] and "PL240.60: Falsely reporting an incident in the first degree" [The State of New York, 2018a]—each carrying prison terms of 5–10 years. Two years and more than a dozen court appearances later the case was finally dismissed in February 2011.

- In 2012, Leigh Van Bryan and Emily Bunting, two UK residents just arriving in Los Angeles for a long-planned holiday, were detained in Customs and locked up for 12 h in a cell for interrogation [Compton, 2012]. Van Bryan's name had been placed on a "One Day Lookout" list maintained by Homeland Security for "intending to come to the US to commit a crime," while Bunting was charged for traveling with him. The source of this were two tweets Van Bryan had made several weeks before his departure. The first read *"3 weeks today, we're totally in LA pissing people off on Hollywood Blvd and diggin' Marilyn Monroe up!"*—according to Van Bryan a quote from his favorite TV show "Family Guy." The second tweet read *"@MelissaxWalton free this week, for quick gossip/prep before I go and destroy America?"* Despite explaining that "destroying" was British slang for "party," both were denied entry and put on the next plane back to the UK. Both were also told that they had been removed from the customary Visa Waiver program that is in place for most European passport holders and instead had to apply for visas from the U.S. Embassy in London before ever flying to the U.S. again [Hartley-Parkinson, 2012].

In both cases, posts on social media that were not necessarily secret, yet implicitly assumed to be for friends only, ended up being picked up by law enforcement, who did not appreciate

---

[22]See https://www.signal.org/.

[23]See https://www.whatsapp.com/. Note that since 2016, WhatsApp supports the encryption of all information being exchanged, though the metadata, i.e., who is chatting with whom and when, is still available to WhatApp's owner, Facebook.

the "playful" nature intended by the poster. Did Joe Lipari or Leigh Van Bryan do "something wrong" and hence had "something to hide"? If not, why should they have anything to fear?

"Knowledge is power" goes the old adage, and as these two stories illustrate, one aspect of privacy certainly concerns controlling the spread of information. Those who lose privacy will also lose control over some parts of their lives. In some cases, this is intended. For example, democracies usually require those in power to give up some of their privacy for the purpose of being held accountable, i.e., to control this power. Citizens routinely give up some of their privacy in exchange for law enforcement to keep crime at bay. In a relationship, we usually show our trust in one another by opening up and sharing intimate details, hence giving the other person power over us (as repeatedly witnessed when things turn sour and former friends or lovers start disclosing these details in order to embarrass and humiliate the other).

In an ideal world, we are in control of deciding who knows what about us. Obviously, this control will have limits: your parents ask you to call in regularly to say where you are; your boss might require you to "punch in/out" when you arrive at work and leave, respectively; the tax office may request a full disclosure on your bank accounts in order to compute your taxes; and police can search your house should they have a warrant[24] from a judge.

In the following two sections we look at both sides of the coin: Why do we want privacy, and why might one *not* want it (in certain circumstances)? Some of the motivations for privacy will be distilled from the privacy laws we have seen in the previous section: what do these laws and regulations attempt to provide citizens with? What are the aims of these laws? By spelling out possible reasons for legal protection, we can try to better frame both the values and the limits of privacy. However, many critics argue that too much privacy will make the world a more dangerous place. Privacy should (and does) have limits, and we will thus also look at the arguments of those that think we should have *less* rather than more privacy.

2.2.1   PRIVACY BENEFITS

The fact that so many countries around the world have privacy legislation in place (over 120 countries in 2017 [Greenleaf, 2017]) clearly marks privacy as an important "thing" to protect, it is far from clear to what extent society should support individuals with respect to keeping their privacy. Statements by Scott McNealy, president and CEO of Sun Microsystems,[25] pointing out that *"you have no privacy anyway, get over it"* [Sprenger, 1999], as well as Peter Cochrane's editorial in *Sovereign Magazine* (when he was head of BT[26] Research) claiming that *"all this secrecy is making life harder, more expensive, dangerous and less serendipitous"* [Cochrane, 2000], are representative of a large part of society that questions the point of "too much" secrecy (see our discussion in Section 2.2.2 below).

---

[24]Such a warrant should only be issued based on sufficient evidence ("probably cause").

[25]Sun Microsystems was once a key software and hardware manufacturer of Unix workstations. It was acquired by Oracle in 2009.

[26]BT was formerly called British Telecom, which was the state-owned telecommunication provider in the UK.

In his book *Code and other Laws of Cyberspace* [Lessig, 1999], Harvard law professor Lawrence Lessig tries to discern possible motivations for having privacy[27] in today's laws and social norms. He lists four major driving factors for privacy.

- *Privacy as empowerment:* Seeing privacy mainly as informational privacy, its aim is to give people the power to control the dissemination and spread of information about themselves. A legal discussion surrounding this motivation revolves around the question whether personal information should be seen as a private property [Samuelson, 2000], which would entail the rights to sell all or parts of it as the owner sees fit, or as a "moral right," which would entitle the owner to assert a certain level of control over their data even after they sold it.

- *Privacy as utility:* From the data subject's point of view, privacy can be seen as a utility providing more or less effective protection from nuisances such as unsolicited calls or emails, as well as more serious harms, such as financial harm or even physical harm. This view probably best follows Warren and Brandeis' "The right to be let alone" definition of privacy, where the focus is on reducing the amount of disturbance for the individual, but can also be found, e.g., in U.S. tort law (see Section 2.1.1) or anti-discrimination laws.

- *Privacy as dignity:* Dignity can be described as "the presence of poise and self-respect in one's deportment to a degree that inspires respect" [Pickett, 2002]. This not only entails being free from unsubstantiated suspicions (for example when being the target of a wire tap, where the intrusion is usually not directly perceived as a disturbance), but rather focuses on the *balance* in information available between two people: analogous to having a conversation with a fully dressed person while being naked oneself, any relationship where there is a considerable information imbalance will make it much more difficult for those with less information about the other to keep their poise.

- *Privacy as constraint of power:* Privacy laws and moral norms to that extend can also be seen as a tool for keeping checks and balances on a ruling elite's powers. By limiting information gathering of a certain type, crimes or moral norms pertaining to that type of information cannot be effectively enforced. As Stuntz [1995] puts it: "Just as a law banning the use of contraceptives would tend to encourage bedroom searches, so also would a ban on bedroom searches tend to discourage laws prohibiting contraceptives" (as cited in Lessig [1999]).

Depending upon the respective driving factor, an individual might be more or less willing to give up part of their privacy in exchange for a more secure life, a better job, or a cheaper product. The ability of privacy laws and regulations to influence this interplay between government and citizen, between employer and employee, and between manufacturer or service provider and customer, creates a social tension that requires a careful analysis of the underlying motivations in

---

[27]A similar categorization but centering around *privacy harms* can be found in Joyee De and Le Métayer [2016].

order to balance the protection of the individual and the public good. An example of how a particular motivation can drive public policy is anti-spam legislation enacted both in Europe [European Parliament and Council, 2002] and in the U.S. [Ulbrich, 2003], which provides privacy-as-an-utility by restricting the unsolicited sending of e-mail. In a similar manner, in March 2004 the Bundesverfassungsgericht (the German Supreme Court) ruled that an 1998 amendment to German's basic law enlarging law enforcements access to wire-tapping ("Der Grosse Lauschangriff") was unconstitutional, since it violated human dignity [Der Spiegel, 2004].

This realization that privacy is more than simply providing secrecy for criminals is fundamental to understanding its importance in society. Clarke [2006] lists five broad driving principles for privacy.

- **Philosophical:** A humanistic tradition that values fundamental human rights also recognizes the need to protect an individual's dignity and autonomy. Protecting a person's privacy is inherent in a view that values an individual for their own sake.

- **Psychological:** Westin [1967] points out the *emotional release* function of privacy—moments "off stage" where individuals can be themselves, finding relief from the various roles they play on any given day: *"stern father, loving husband, car-pool comedian, skilled lathe operator, unions steward, water-cooler flirt, and American Legion committee chairman."*

- **Sociological:** Societies do not flourish when they are tightly controlled, as countries such as East Germany have shown. People need room for *"minor non-compliance with social norms"* and to *"give vent to their anger at 'the system,' 'city hall,' 'the boss':"*

  The firm expectation of having privacy for permissible deviations is a distinguishing characteristic of life in a free society [Westin, 1967].

- **Economical:** Clark notes that *"all innovators are, by definition, 'deviant' from the norms of the time,"* hence having private space to experiment is essential for a competitive economy. Similarly, an individual's fear of surveillance—from both private companies and the state—will dampen their enthusiasm in participating in the online economy.

- **Political:** The sociological need for privacy directly translates into political effects if people are not free to think and discuss outside current norms. Having people actively participate in political debate is a cornerstone of a democratic society—a lack of privacy would quickly produce a "chilling effect" that directly undermines this democratic process.

As Clarke [2006] points out, many of today's data protection laws, in particular those drafted around the Fair Information Principles, are far from addressing all of those benefits, and instead rather focus on ensuring that the collected data is correct—not so much as to protect the individual but more so to ensure maximum economic benefits. The idea that privacy is more of an individual right, a right that people should be able to exercise without unnecessary burden, rather than simply an economic necessity (e.g., to make sure collected data is correct), is a relatively recent development. Representative for this paradigm shift was the so-called "census-verdict" of the German federal constitutional court (Bundesverfassungsgericht) in 1983, which extended the existing *right to privacy of the individual* (Persönlichkeitsrecht) with the *right of self-determination over personal data* (informationelle Selbstbestimmung) [Mayer-Schönberger, 1998].[28] The judgment reads as follows.[29]

> If one cannot with sufficient surety be aware of the personal information about oneself that is known in certain part of his social environment, …can be seriously inhibited in one's freedom of self-determined planning and deciding. A society in which the individual citizen would not be able to find out who knows what when about them, would not be reconcilable with the right of self-determination over personal data. Those who are unsure if differing attitudes and actions are ubiquitously noted and permanently stored, processed, or distributed, will try not to stand out with their behavior. …This would not only limit the chances for individual development, but also affect public welfare, since self-determination is an essential requirement for a democratic society that is built on the participatory powers of its citizens [Reissenberger, 2004].

The then-president of the federal constitutional court, Ernst Benda, summarized his private thoughts regarding their decision as follows.[30]

> The problem is the possibility of technology taking on a life of its own, so that the actuality and inevitability of technology creates a dictatorship. Not a dictatorship of people over people with the help of technology, but a dictatorship of technology over people [Reissenberger, 2004].

The concept of self-determination over personal data[31] constitutes an important part of European privacy legislation with respect to ensuring the autonomy of the individual. First, it extends the Fair Information Principles with a participatory approach, which would allow the individual to decide beyond a "take it or leave it" choice over the collection and use of his or her personal information. Second, it frames privacy protection no longer only as an individual right, but emphasizes its positive societal and political role. Privacy not as an individual fancy, but as an obligation of a democratic society, as Julie Cohen notes:

> Prevailing market-based approaches to data privacy policy …treat preferences for informational privacy as a matter of individual taste, entitled to no more (and often much less) weight than preferences for black shoes over brown, or red wine over

---

[28]The finding was triggered by the controversy surrounding the national census announcement on April 27, 1983, which chose the unfortunate wording "Totalzählung" and thus resulted in more than hundred constitutional appeals (Verfassungsbeschwerde) to the federal constitutional court [Reissenberger, 2004].

[29]Translation by the authors.

[30]Translation by the authors.

[31]Often abbreviated to *data self-determination*.

white. But the values of informational privacy are far more fundamental. A degree of freedom from scrutiny and categorization by others promotes important non-instrumental values, and serves vital individual and collective ends [Cohen, 2000].

The GDPR additionally includes a number of protection mechanisms that are designed to strengthen the usually weak bargaining position of the individual. For example, Article 9 of the GDPR specifically restricts the processing sensitive information, such as ethnicity, religious beliefs, political or philosophical views, union membership, sexual orientation, and health, unless for medical reasons or with the explicit consent of the data subject.

2.2.2   LIMITS OF PRIVACY

There are certainly limits to an individual's privacy. While the GDPR obliges EU member states to offer strong privacy protection, many of them at the same time keep highly detailed records on their citizens, both in the interest of fighting crime but also in order to provide social welfare and other civil services.[32]

A good example for this tension between the public good and the protection of the individual can be found in the concept of communitarianism. Communitarians like Amitai Etzioni, professor for sociology at the George Washington University in Washington, D.C., and founder of the Communitarian Network, constantly question the usefulness of restricting society's power over the individual through privacy laws, or more general, to "articulate a middle way between the politics of radical individualism and excessive stateism" [Etzioni, 1999].

In his 1999 work *The Limits of Privacy* [Etzioni, 1999], Etzioni gives the example of seven-year-old Megan Kanka, who in 1994 was raped and strangled by her neighbor Jesse Timmend-equas. No one in the neighborhood knew at that time that Timmendequas had been tried and convicted of two prior sex offenses before, and had served six years in prison for this just prior to moving in next to the Kankas. Megan Kanka's case triggered a wave of protests in many U.S. states, leading to virtually all states implementing some sort of registration law for convicted sex offenders, collectively known as "Megan's Law." Depending on the individual state, such registration procedures range from registering with the local police station upon moving to a new place, to leaving blood and saliva samples or even having to post signs in one's front yard reading "Here lives a convicted sex offender"[33] [Solove and Rotenberg, 2003].

While many criticize Megan's Law for punishing a person twice for the same crime (after all, the prison sentence has been served by then—the perpetual registration requirement equals a lifelong sentence and thus contradicts the aim of re-socialization), others would like even more rigorous surveillance (e.g., with the help of location-tracking ankle bracelets) or even a

---

[32]In its Article 23, the GDPR allows member states to limit the applicability of the Regulation in order to safeguard, e.g., national or public security.

[33]In May 2001, a judge in Texas ordered 21 convicted sex offenders not only to post signs in their front yards, but also place bumper stickers on their cars stating: "Danger! Registered Sex Offender in Vehicle" [Solove and Rotenberg, 2003].

lifelong imprisonment in order to prevent any repeated offenses.[34] A similar lifelong-custody mechanism passed in 2004 a public referendum in Switzerland: before being released from their prison sentence, psychologists will have to assess a sex offender's likelihood for relapse. Those with a negative outlook will then be taken directly into lifelong custody.

But it is not only violent crimes and homeland security that makes people wonder whether the effort spent on protecting personal privacy is worth it. Especially mundane everyday data, such as shopping lists or one's current location—things that usually manifest themselves in public (in contrast to, say, one's diary, or one's bank account balance and transactions)—seem to have no reason for protection whatsoever. In many cases, collecting such data means added convenience, increased savings, or better service for the individual: using detailed consumer shopping profiles, stores will be able to offer special discounts, send only advertisements for items that really interest a particular customer, and provide additional information that is actually relevant to an individual. And, as Lessig remarks, any such data collection is not really about any individual at all: "[N]o one spends money collecting these data to actually learn anything about you. They want to learn about people *like* you" [Lessig, 1999].

What could be some of the often cited dangers of a transparent society then? What would be the harm if stores had comprehensive *profiles* on each of their customers in order to provide them with better services?

One potential drawback of more effective advertisement is the potential for manipulation: if, for example, one is identified as a mother of teenagers who regularly buys a certain breakfast cereal, a targeted advertisement to buy a competitor's brand at half the price (or with twice as many loyalty points) might win the kid's favor, thus prompting the mother to switch to the potentially more expensive product (with a higher profit margin). A similar example of "effective advertising" can be found in the Cambridge Analytica scandal of 2018 [Lee, 2018, Meyer, 2018], which saw a political data firm harvest the private profiles of over 50 million Facebook users (mostly without their knowledge) in order to create "psychographic" profiles that were then sold to several political campaigns (the 2016 Trump campaign, the Brexit "Leave" campaign) in order to target online ads. Presumably, such information allowed those campaigns to identify voters most open to their respective messages. Profiles allow a process that sociologist David Lyon calls *social sorting* [Lyon, 2002]:

> The increasingly automated discriminatory mechanisms for risk profiling and social categorizing represent a key means of reproducing and reinforcing social, economic, and cultural divisions in informational societies [Lyon, 2001].

This has implication on both the individual and societal level. For democratic societies, a thoroughly profiled population exposed to highly target political ads may become increasingly divided [The Economist, 2016]. On an individual level, the benefits of profiling would depend on the existing economic and social status. For example, since a small percentage of customers

---

[34]Another problem with this approach is its broad application toward any "sex-offenses:" in some states, this also puts adult homosexuals or underage heterosexual teenagers having consensual sex on such lists.

(whether it be in supermarkets or when selling airline tickets) typically makes a large percent-age of profits,[35] using consumer loyalty cards or frequent flyer miles would allow vendors to more accurately determine whether a certain customer is worth fighting for, e.g., when having to decide if a consumer complaint should receive fair treatment.

This might not only lead to withholding information from customers based on their pro-files, but also to holding this information *against* them, as the example of Ron Rivera in Chap-ter 1 showed. In a similar incident, a husband's preference for expensive wine that was well documented in his supermarket profile, allowed his wife to claim a higher alimony after having subpoenaed the profile in court. Even if such examples pale in comparison to the huge number of transactions recorded everyday worldwide, they nevertheless indicate that this massive col-lection of mundane everyday facts will further increase through the use of mobile and pervasive computing, ultimately adding a significant burden to our lives, as Lessig explains:

> The burden is on you, the monitored, first to establish your innocence, and sec-ond, to assure all who might see these ambiguous facts, that you are innocent [Lessig, 1999].

This silent reversal of the classical presumption of innocence can lead to significant dis-advantages for the data subject, as the examples of comedian Joe Lipari (page 23), UK-couple Leigh Van Bryan and Emily Bunting (page 23), and firefighter Philip Scott Lyons (page 1) have shown. Another example for the sudden significance of these profiles is the fact that shortly af-ter the September 11 attacks, FBI agents began collecting the shopping profiles and credit card records of each of the suspected terrorists in order to assemble a terrorist profile [Baard, 2002].[36] First reports of citizens who were falsely accused, e.g., because they shared a common name with a known terrorist [Wired News] or had a similar fingerprint [Leyden, 2004], illustrate how dif-ficult it can be for an individual to contest findings from computerized investigative tools.

Complete transparency, however, may also help curb governmental power substantially, according to David Brin, author of the book "The Transparent Society" [Brin, 1998]. In his book, Brin argues that losing our privacy can ultimately also have advantages: While up to now, only the rich and powerful had been able to spy on common citizens at will, the next technology would enable even ordinary individuals to "spy back," to "watch the watchers" in a society without secrets, where everybody's actions could be inspected by anybody else and thus could be held accountable, where the "surveillance" from above could now be counteracted by "sousveillance" from below [Mann et al., 2003].

Critics of Brin point out that "accountability" is a construct defined by public norms and thus will ultimately lead to a homogenization of society, where the moral values of the majority

---

[35]The Guardian cites IBM-analyst Merlin Stone with saying "In every sector, the top 20% of customers give 80% of the profit" [Guardian].

[36]Interestingly enough, the main shopping characteristic for all of the suspected terrorists wasn't a preference for Middle-eastern food, but rather a tendency to order home-delivery pizza and paying for it by credit card.

will threaten the plurality of values that forms an integral part of any democracy, simply by holding anybody outside of the norm "accountable" [Lessig, 1999].

The ideal level of privacy can thus have very different realities, depending on what is technically feasible and socially desirable. The issues raised above are as follows.

1. *Communitarian:* Personal privacy needs to be curbed for the greater good of society (trusting the government). Democratic societies may choose to appoint trusted entities to oversee certain private matters in order to improve life for the majority.

2. *Convenience:* The advantages of free flow of information outweighs the personal risks in most cases. Only highly sensitive information, like sexual orientation, religion, etc. might be worth protecting. Semi-public information like shopping habits, preferences, contact information, and even health information, might better be publicly known so that one can enjoy the best service and protection possible.

3. *Egalitarian:* If everybody has access to the same information, it ceases to be a weapon in the hands of a few well-informed. Only when the watchers are being watched, all information they hold about an individual is equally worth the information the individual holds about them. Eventually, new forms of social interaction will evolve that are built upon these symmetrical information assets.

4. *Feasibility:* What can technology achieve (or better: prevent)? All laws and legislation require enforceability. If privacy violations are not traceable, the much stressed point of accountability (as developed in the fair information practices) becomes moot.

2.3   CONCEPTUALIZING PRIVACY

The prior sections should already have helped to illustrate some of the many different ways of understanding what privacy is (or should be). In this section, we will present some of the many attempts to "capture" the nature of privacy in definitions and conceptual models.

2.3.1   PRIVACY TYPES

The historic overview in Section 2.1.1 provides a sense of how the understanding of privacy has changed over the years—continuously adding "things to protect," often in response to novel technological developments. Clearly, back in the late 19th century, with no computerized data processing around, privacy of "data" was not much of an issue. Instead, as evident in Warren and Brandeis' work or William Pitt's Excise Bill speech ("The poorest man may in his cottage bid defiance to all the forces of the Crown...") [Brougham, 1839], the protection of the home—*territorial privacy*—was the most prevalent aspect of privacy protection. While this conception dates back to the 18th century, defining privacy as "protected spaces" is still relevant today. For example, workplace privacy—which is rooted in the concept of territorial privacy—sees renewed interest, given how fewer and fewer people have an actual desk in an office that would mark a defined "territory," but instead use hot desking[37] or work in coffee shops or co-working spaces. What exactly is the "territory" that one should protect here?

With the advent of the telegraph and the telephone, *communication privacy* became another facet of privacy concerns. While letters had long since enjoyed some sort of privacy guarantees—either by running a trusted network of private messengers (e.g., religious orders), or by "corporate" guarantees from early postal companies such as the Thurn-and-Taxis Post [Schouberechts, 2016]—these new forms of remote communication quickly required legal provisions for safeguarding their contents. Email and recently instant messengers and Voice-over-IP have again made communication privacy a timely issue.

Maybe furthest back dates the idea of *bodily privacy*, visible in the earliest "privacy laws" (which were not called privacy laws at the time) against peeping toms [Laurant, 2003]. Today, bodily privacy remains relevant as, e.g., international travelers may be subject to strip searches at airports in countries with strict drug laws, or workers may be forced to accept mandatory drug tests by their employer.

Today's most prevalent "privacy type" comes from the 1960s, when automated data processing first took place on a national scale. Alan Westin, then professor of public law and government at Columbia University, defined privacy in his groundbreaking book *Privacy and Freedom* as "*the claim of individuals, groups, or institutions to determine for themselves when, how, and to what extent information about them is communicated to others*" [Westin, 1967]. The "thing" to protect here is a person's "data"—information about them that, once collected, may be shared with others without their knowledge or consent. Not surprisingly, Westin's type of privacy is called *information privacy* or sometimes also *data privacy*.

Figure 2.2 illustrates how each of these privacy types relates to a different aspect (in center of figure): bodily privacy to the person, territorial privacy to physical space, communication privacy to our social interactions, and information privacy to stored "files" about us. Figure 2.2 also shows how, as technology moves forward, these privacy types are being challenged anew. For example, Westin's "information privacy" is being challenged by today's online profiles: with each website we visit being connected to countless ad networks, content delivery networks, performance trackers, affiliate sites, and other third parties a single page view can easily result in dozens if not hundreds of traces being left in databases around the world—practically impossible for the individual to keep track of, lest control. Similarly, communication privacy has long ceased to be an issue that pertains only to postal services: email, chat, mobile messaging, and online social networking potentially allow both governments and companies to observe our communication patterns in a more fine-grained fashion than ever before. And by using a modern smartphone for our communication with others, we also provide countless of companies, if not governments, with detailed information on our whereabouts and activities. Now, DNA analysis is becoming

---

[37]Hot desking describes the practice of office workers sharing a single physical desk/workplace, often on a first come–first use basis. Personal belongings are kept in movable (and lockable) containers that can easily be moved to another empty desk the next morning.

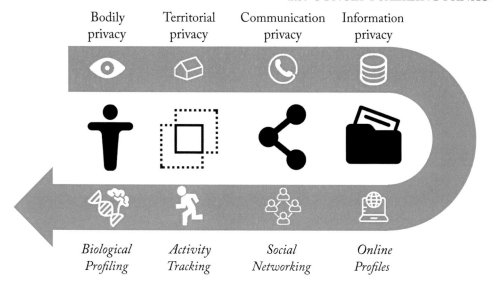

Figure 2.2: *Core Privacy Types and Today's Challenges.* Across history, privacy concerns shifted as technology made things possible that were not possible before. The four broad types of privacy—bodily, territorial, communication, and information—are being challenged by new technology: online profiling, social networking, activity tracking, and biological profiling.

affordable and commonplace,[38] which means that biological profiling will challenge our bodily privacy like never before.

2.3.2 PRIVACY CONSTITUENTS

The early characterization of privacy as "the right to be let alone" by Warren and Brandeis [1890] clearly captured only a narrow part[39] of the complex privacy problem. Westin [1967] expanded this concept of "privacy as solitude" to capture a wider range of settings that all can be motivated with the need for privacy, stressing the fact that *the individual's desire for privacy is never absolute, since participation in society is an equally powerful desire* [Westin, 1967]. Westin enumerated four distinct "privacy states" (see Figure 2.3) that all describe different forms of privacy: beyond the idea of "privacy as solitude," Westin describes "intimacy"—the state of sharing intimate information with another person, "reserve"—the act of "standing apart," e.g., at a party, in order to (even if only temporarily) disengage from others ("the creation of a psychological

---

[38]Companies like AncestryDNA (www.ancestry.com/dna/) and 23andMe (www.23andme.com) already sell kits for less than USD $100 that allow consumers to get an in-depth analysis of their DNA. They typically reserve the right to "share aggregate information about users genomes to third parties" [Seife, 2013].

[39]Even within such a narrow definition, however, the inherent conceptual model can become quite complex, as the various privacy torts derived from it (protection from unwanted intrusion, public disclosure, adverse publicity, and appropriation) illustrate.

barrier against unwanted intrusion"), and "anonymity"—blending in with a crowd so as to be non-distiguishable from others ("the individual is in public places or performing public acts but still seeks, and finds, freedom from identification and surveillance"). Each of these states offer some sort of privacy, even though they involve very different physical and social settings.

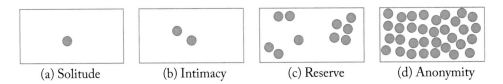

    (a) Solitude   (b) Intimacy   (c) Reserve   (d) Anonymity

Figure 2.3: *Westin's Privacy States* based on Westin [1967] defines four privacy states, or "experiences:" Solitude, Intimacy, Reserve, and Anonymity.

Solitude is the "positive" version of loneliness, the act of "being alone without being lonely." Solitude plays an important role in psychological well-being, offering benefits such as freedom, creativity, and spirituality [Long and Averill, 2003].

Intimacy is probably a concept as complex as privacy, as it may refer to "feelings, to verbal and nonverbal communication processes, to behaviors, to people's arrangements in space, to personality traits, to sexual activities, and to kinds of long-term relationships" [Reis et al., 1988]. Yet, it is clear that intimacy is an essential component of forming the types of close relationships [Levinger and Raush, 1977] that are essential to our psychological well-being [Baumeister and Leary, 1995]. Gerstein [Gerstein, 1978] argues that "intimacy simply could not exist unless people had the opportunity for privacy." Paradoxically, Nisenbaum [1984] finds that even solitude often helps create intimacy, by prompting "feelings of connection with another person" [Long and Averill, 2003].

Anonymity provides us with what Rössler [2001] calls *decisional privacy*: "*securing the interpretational powers over one's life.*" The freedom to decide for oneself "*who do I want to live with; which job to take; but also: what clothes do I want to wear.*" Anonymity thus helps to ensure the *autonomy* of the individual, protecting one's independence in making choices central to personhood.

The interplay between solitude, intimacy, and anonymity (and hence autonomy) ultimately shapes our *identity*. Westin describes this as follows.

> Each person is aware of the gap between what he wants to be and what he actually is, between what the world sees of him and what he knows to be his much more complex reality. In addition, there are aspects of himself that the individual does not fully understand but is slowly exploring and shaping as he develops [Westin, 1967].

According to Arendt [1958], privacy is essential for developing an individual identity because it facilitates psychological and social depth, and protects aspects of a person's identity that cannot withstand constant public scrutiny—be it publicly chastised preferences and practices, or silly tendencies and other behavior people self-censor when being observed.

Autonomy allows us to be what we want to be; intimacy and solitude help us to explore and shape our "complex reality," both through the intimate exchange with others. This not only connects with the previously-mentioned *emotional release* function of privacy, but also with what Westin [1967] calls the "safety-value" function of privacy, e.g., the *"minor non-compliance with social norms"* and to *"give vent to their anger at 'the system,' 'city hall,' 'the boss'"*:

> The firm expectation of having privacy for permissible deviations is a distinguishing characteristic of life in a free society [Westin, 1967].

In that sense, privacy protects against extrinsic and intrinsic losses of freedom [Nissenbaum, 2009, p. 75]. Nissenbaum argues that *"privacy is important because it protects the diversity of personal choices and actions, not because it protects the freedom to harm others and commit crimes"* [Nissenbaum, 2009, p. 77].

Other privacy constituents can be drawn from its *implementation*, i.e., how current social, legal, and technical solutions attempt to provide privacy to citizens, customers, employees, or users. We can roughly group these approaches into three categories: secrecy, transparency, and control.

- *Secrecy* is often equated with privacy, in particular in security scholarship, where privacy is simply another word for *confidentiality*. At the outset, it seems as if privacy without secrecy does not make sense: if others know our information, we have lost our privacy, surely? Such thinking implies a binary nature of privacy: we either have it or do not have it, based on the knowledge of others. Similar to Warren and Brandeis' conception of privacy as "the right to be let alone," such a binary view is neither realistic nor practical. When I confide in a trusted friend something in private, I surely do not expect my disclosure to invalidate my privacy. I instead expect my friend to hold this information in confidence, making it a shared secret between the two of us, rather than public information. I expect my doctor, who knows a lot about my health, to keep this information private—in fact, many professionals are often bound by law to keep private information of others secret (e.g., lawyers, clerics). Note that in these cases, secrecy does not entirely vanish, it just includes more people who share the same confidential information. An interesting corner case of secrecy is anonymity, often also called *unlinkability*. While confidentiality makes it hard for others to find out a certain secret information, unlinkability removes its connection to a person. As it is the case with confidentiality, unlinkability is not a binary value but comes in many different shades. *Data minimization* is the bridge between the two: It aims to ensure that only those data elements are collected that are essential for a particular purpose—all non-essential information is simply not collected.

- *Transparency* can be seen as the flip side of secrecy. If secrecy prevents others from knowing something about me, transparency allows me to know what others know about me. In its simplest form, transparency requires *notice*: informing someone about what data is being collected about them, or why this information is being collected, or what is already known about them. Transparency is often the most basic privacy requirement, as it thwarts secret record keeping. While individuals may have no say about their data being collected, they may at least understand that their information is being recorded and act accordingly (and thus retain their autonomy). Obviously, transparency by itself does not mean that one's privacy is being protected—the fact that one spots a nosy paparazzi taking photos from afar does not help much once one's secret wedding pictures appear in a tabloid paper.

- *Control* is often the key ingredient that links secrecy and transparency: if people can freely control "when, how, and to what extend information about them is communicated to others" [Westin, 1967], they can decide who they want to take into their confidence and when they like to keep information private or even remain anonymous. A practical (and minimal) form of control is "consent," i.e., a person's affirmative acknowledgment of a particular data collection or data processing practice. Consent may be implicit (e.g., by posting a sign "video surveillance in progress at this premise" by the door through which visitors enter) or explicit (e.g., by not starting an online data collection practice unless a person has checked a box in the interface). Implicit and explicit consent are thus seen as giving individuals a *choice*: if they do not want their data collected, the have the option of not selecting a particular option, or simply not proceeding beyond a particular point (either physical or in a user interface). Together, notice and choice offer what Solove [2013] calls "privacy self-management." Such control tools form the basis for many privacy laws worldwide, as we discussed in Section 2.1. While control seems like the ideal "fix" for enabling privacy, the feasibility of individuals effectively exercising such control is questionable. Solove [2013] notes that individuals are often ill-placed to make privacy choices: "privacy decisions are particularly susceptible to problems such as bounded rationality, the availability heuristic, and framing effects because privacy is so complex, contextual, and difficult to conceptualize." In most cases, people simply "lack enough background knowledge to make an informed choice," or "there are simply too many entities that collect, use, and disclose people's data for the rational person to handle."

With these constituents in mind, Gavison [1984] goes on to define privacy as being comprised of "solitude, anonymity, and control," while Simmel [1968] puts it similarly, yet expanding somewhat on Gavinson.

> Privacy is a concept related to solitude, secrecy, and autonomy, but it is not synonymous with these terms; for beyond the purely descriptive aspects of privacy as isolation from the company, the curiosity, and the influence of others, privacy implies a normative element: the right to exclusive control to access to private realms [Simmel, 1968].

Contrary to Westin and Rössler, Gavinson and Simmel describe privacy not as an independent notion, but rather as an amalgam of a number of well established concepts, something that constitutes itself only through a combination of a range of factors. While Westin also relates privacy to concepts such as solitude, group seclusion, anonymity, and reserve [Cate, 1997], he calls them *privacy states*, indicating that these are merely different sides to the same coin.

2.3.3 PRIVACY EXPECTATIONS

The boundaries implicit in the privacy states described by Westin [1967] are a key part of modeling privacy *expectations*. As is also the case for security, privacy is not a goal in itself, not a service that people want to subscribe to, but rather an expectation of being in a state of protection without having to actively pursue it. All else being equal, users undoubtedly would prefer systems without passwords or similar access control mechanisms, as long as they would not suffer any disadvantages from this. Only if any of their files are maliciously deleted or illegally copied, users will regret not having any security precautions in place. So what would be the analogy to a "break-in" from a privacy point of view?

Marx [2001] identified *personal border crossings* as a core concept for understanding privacy: *"Central to our acceptance or sense of outrage with respect to surveillance ...are the implications for crossing personal borders."* Marx differentiates between four such border crossings that are perceived as privacy violations.

- *Natural borders:* Physical limitations of observations, such as walls and doors, clothing, darkness, but also sealed letters, telephone calls. Even facial expressions can form a natural border against the true feelings of a person.

- *Social borders:* Expectations about confidentiality for members of certain social roles, such as family members, doctors, or lawyers. This also includes expectations that your colleagues will not read personal correspondance addressed to you, or material that you left lying around the photocopy machine.

- *Spatial or temporal borders:* The usual expectations of people that parts of their life, both in time and social space, can remain separated from each other. This would include a wild adolescent time that should not interfere with today's life as a father of four, or different social groups, such as your work colleagues and friends in your favorite bar.

- *Borders due to ephemeral or transitory effects:* This describes what is best known as a "fleeting moment," an unreflected utterance or action that we hope gets forgotten soon, or old pictures and letters that we put out in our trash. Seeing audio or video recordings of such events later, or observing someone sifting through our trash, will violate our expectations of being able to have information simply pass away unnoticed or forgotten.

Whenever personal information crosses any of these borders without our knowledge, our potential for possible actions—our decisional privacy—is affected. When someone at the office suddenly mentions family problems that one has at home, or if circumstances of our youth suddenly are being brought up again even though we assumed that they were long forgotten, we perceive a violation of our local, informational, or communication privacy. This violation is by no means an absolute measure, but instead depends greatly on the individual circumstances, such as the kind of information transgressed, or the specific situation under which the information has been disclosed. The effects such border crossing have on our lives, as well as the chances that they actually happen, are therefore a highly individual assertion.

Similarly, Nissenbaum [1998] investigated expectations of privacy in public situations, e.g., private conversations in a public restaurant, ultimately proposing a framework of *contextual integrity* [Nissenbaum, 2004, 2009] that defines privacy along contextualized privacy expectations, rather than a private/public dichotomy. To Nissenbaum, privacy expectations relate to context-dependent norms of information flows that are characterized by four key aspects: contexts, actors, attributes, and transmission principles [Nissenbaum, 2009].

- *Contexts:* Contexts describe the general institutional and social circumstances (e.g., healthcare, education, family, religion, etc.) in which information technology is used or information exchange takes place. Contexts also include the activities in which actors (in different roles) engage, as well as the purposes and goals of those activities (Nissenbaum calls this *values*). Contexts and associated informational norms can be strictly specified or only sparsely and incompletely defined—for example, the procedure of voting vs. an informal business meeting. Expectations of confidentiality are clearly defined in the first place, but less clear in the second case. People often engage in multiple contexts at the same time which can be associated with different, potentially conflicting informational norms. For instance, talking about private matters at work.

- *Actors:* Actors are senders, receivers, and information subjects who participate in activities. Actors fill specific roles and capacities depending on the contexts. Roles define relationships between various actors, which express themselves through the level of intimacy, expectations of confidentiality, and power dynamics between actors. Informational norms regulate information flow between actors.

- *Attributes:* Attributes describe the type and nature of the information being collected, transmitted, and processed. Informational norms render certain attributes appropriate or inappropriate in certain contexts. The concept of appropriateness serves to describe what are acceptable actions and information practices.

- *Transmission principles:* Transmission principles constrain the flow of information between entities. They are associated with specific expectations. Typical transmission principles are confidentiality, reciprocity or fair exchange of information, and whether an actor deserves or is entitled to receive information.

Norms of information flows can be either implicitly understood or explicitly codified. Common types of norms are moral norms, conventions of etiquette, rules, and procedures. In order to identify the potential of a novel technology to increase the danger of privacy violations, Nissenbaum proposes a multi-step process [Nissenbaum, 2009, p. 148ff, 182ff], similar to a privacy impact assessment (see, e.g., Wright and De Hert [2012]). Her process starts by establishing the social context in which a technology is used, the key actors, the affected attributes, i.e., the involved information, and the principles of transmission. Contextual integrity and privacy expectations are violated if the introduction of an information technology or practice changes any of these aspects. Furthermore, one should analyze how the technology affects moral and political factors, e.g., power structures, fairness, or social hierarchies. Identified threats should then be further analyzed for their impact on goals and values in the specific context. Based on these assessments, contextual integrity recommends either for or against the information practice.

Altman [1975] proposes a corresponding process-oriented view on privacy in his *privacy regulation theory*. He defines privacy as *"the selective control of access to the self or to one's group"* and argues that individuals adjust their privacy based on internal and external changes. Internal changes can be caused by changes in personal preference, past experiences, or new knowledge. External changes pertain to changes in the environment and the context. According to Altman, privacy is a dynamic, dialectic, and non-monotonic process. An individual regulates privacy and social interaction through adjusting own *outputs* as well as *inputs* by others. Outputs roughly correspond to information that is being disclosed or observable; inputs to potential for invasions and disturbances. In social interaction, adjustments rely on *verbal, paraverbal*, and *nonverbal behavioral mechanisms*, such as revealing or omitting information (verbal), changing intonation and speaking volume (paraverbal), or using posture and gestures to non-verbally express and control personal space and territory. Schwartz [1968] analyzes how entering and leaving personal spaces, and even the use of doors, is governed by rules of appropriateness and privacy expectations.

A critical part of Altman's theory is the distinction between *desired privacy* and *achieved privacy*. Desired privacy is defined by an individual's privacy preferences and privacy expectations. Achieved privacy is the actual privacy level obtained or achievable in a given situation with the means for privacy control available in that situation. If achieved privacy is lower than desired privacy, privacy expectations are violated and the individual feels exposed. Achieving more privacy than desired causes *social isolation*. Thus, the privacy regulation process aims for an optimal privacy level in which desired and achieved privacy are aligned.

2.3.4   A PRIVACY TAXONOMY

Solove [2008] argues that striving for a singular definition for all privacy characteristics may be misguided due to the diversity of the topic. Furthermore, information alone may not be a sufficient indicator for associated privacy expectations, because the sensitivity of information depends on the purposes for which data subjects want to conceal it or other parties want to use it. Thus, Solove argues for a more pragmatic, contextualized, and pluralistic view on privacy. Privacy problems disrupt particular activities. Whenever a privacy problem surfaces in any given situation, a corresponding privacy interest must exist. As a consequence, Solove suggests to define privacy through its disruptions and issues in specific contextual situations rather than general core characteristics. Privacy protection mechanisms and regulations have to address multiple interconnected problems and balance conflicting interests. In order to map the topology of these interconnected problems, and support the identification of compromises that protect both privacy and the conflicting interest, Solove proposes a comprehensive privacy taxonomy structured by generalized privacy problems [Solove, 2006, 2008].

Solove divides privacy problems into four major categories, as shown in Figure 2.4, loosely following an information flow from the data subject to data controllers and further on to potential third parties. Each of these categories is further divided into subcategories.

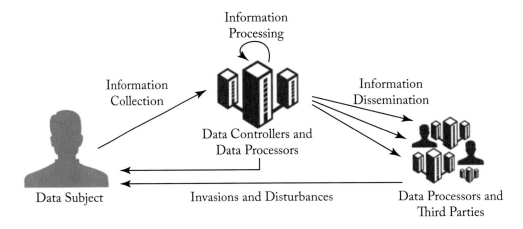

Figure 2.4: Four categories of privacy issues based on Solove's privacy taxonomy based on Solove [2008].

Information Collection

Solove distinguishes two types of information collection: *surveillance* and *interrogation*. Surveillance is the passive observation of the data subject by others. He argues that while not all observations disrupt privacy, continuous monitoring does. Surveillance disrupts privacy because people may feel anxious and uncomfortable and even alter their behavior when they are being watched. Covert surveillance has the additional effect that it creates a power imbalance because the data subject can be observed without being able to see the observer.

Bentham's Panopticon purposefully leverages this effect in the architectural design of a prison [Bentham, 1787]. Bentham designed the panopticon as a circular prison with cells on the outside that are open toward the middle. A guard tower at the center is fitted with small window slits facing in all directions. Thus, guards could watch any prisoner at any time, while inmates would not know when they are actually being watched.

In a less Orwellian sense, surveillance also includes privacy issues caused by incidental observations, such as private information visible on a screen or someone observing a typed password, also known as shoulder surfing [Schaub et al., 2012]. Hawkey and Inkpen [2006] investigate the dimensions of incidental information privacy.

In contrast, interrogation constitutes active information collection. The data subject is directly exposed to an inquisitive party, which may pressure the data subject to disclose details. Similar to surveillance, less evocative interrogation issues also occur in common situations. For example, when a questionnaire or a registration form asks for more information than required, or when social pressure leads to revealing information one would have kept private otherwise.

Information Processing

Solove's second category, information processing, contains five potential privacy harms, which all occur after information has been collected, and therefore without direct involvement of the data subject: *aggregation, identification, insecurity, secondary use,* and *exclusion.*

Aggregation of information about one person facilitates profiling. While such aggregation can have benefits, it often violates the data subjects' expectations in terms of what others should be able to find out about them. However, the effects of aggregation are typically less direct, because the data has already been collected previously. The main issue is that multiple innocuous pieces of information gain privacy sensitivity when combined.

Identification is the process of linking some information to a specific individual, sometimes also called *re-identification* or *de-anonymization.* Presumably anonymized data may contain sufficient information to link the provided information back to the individual. For instance, Sweeney [2002] showed that zip code, gender, and date of birth provided in U.S. census data are sufficient to uniquely identify 87% of the U.S. population. The risk of de-anonymization has also been demonstrated in the context of presumably anonymized medical records [Benitez and Malin, 2010], genome research data [Malin and Sweeney, 2004], location traces [Gruteser and Hoh, 2005], and home/work location pairs [Golle and Partridge, 2009].

The lack of proper security ("insecurity") of data processing and stored data is also a significant privacy risk, because it facilitates identity theft and distortion when information about an individual is more readily accessible by unauthorized entities than it should be. Therefore, all discussed data protection frameworks place a strong emphasis on the security of collected personal data.

The term "secondary use" describes any use of collected information beyond the purposes for which it was collected. As the data subject did not consent to secondary use by definition, it violates privacy expectations. The main issue here is that data subjects may have provided different information if they would have been aware of the secondary use.

Solove calls the lack of appropriate means for data subjects to learn about the existence of collected personal data, to access those data, and to rectify it "exclusion." Exclusion runs contrary to the data protection principles of participation and transparency.

Information Dissemination

Solove's third category, information dissemination, summarizes seven privacy issues that concern the further disclosure or spread of personal information.

Breach of confidentiality is the violation of trust in a specific relationship by revealing secret information associated with that relationship. *Disclosure* is the dissemination of true information about a data subject without consent. It violates the data subject's information self-determination. Disclosure can adversely affect the data subject's reputation. *Distortion* is similar to disclosure with the difference that false or misleading information about a person is being willfully disseminated, often with the intention of harming that person's reputation.

Exposure is very similar to disclosure, but Solove notes that it pertains to revealing physical or emotional attributes about a person, such as information about the person's body and health. Thus, exposure violates bodily privacy and affects the person's dignity rather than reputation.

Increased accessibility does not directly disclose information to any specific party but makes it generally easier to access aggregated information about an individual. Although information might have been previously publicly available, aggregation and increased accessibility increase the risk of actual disclosure.

Blackmail is the threat to expose private information if the blackmailer's demands are not met. Blackmail is a threat of disclosure enabled by a power imbalance created by information obtained by the blackmailer. *Appropriation*, on the other hand, is the utilization of another person's identity for one's own benefit. It is sometimes also referred to as exploitation.

Invasion

Solove's fourth category is concerned with privacy *invasion*, featuring two privacy issues: *intrusion* and *decisional interference*. While Solove's other three categories mainly deal with information privacy, invasion does not involve personal information directly.

Intrusion is the violation of someone's personal territory, however that territory may be defined. One can intrude in someone's physical territory or private space, but intrusion can also pertain to disrupting private affairs. Such *"realms of exclusion"* [Solove, 2008] facilitate interaction with specific people without interference, and also exist in otherwise public environments, e.g., having a private conversation at a restaurant.

Decisional interference is a privacy issue where governmental regulations interfere with the freedom of personal decisions and self-determination (Rössler's concept of *decisional privacy*), e.g., in the case of sexuality or religious practices. Solove argues that these are not merely issues of autonomy but are strongly associated with information privacy. The risk of potential disclosure can severely inhibit certain decisions of an individual.

Solove's taxonomy provides a comprehensive framework to reason about types of privacy violations. His goal was to facilitate the categorization of privacy violations in specific cases to obtain appropriate legal regulations and rulings. He ignores individual privacy preferences in his taxonomy on purpose, because, according to him, it is virtually impossible to protect individual, varying privacy expectations in a legal framework [Solove, 2008, p. 70]. However, he recognizes that personal privacy preferences play an important role in shaping individual expectations of privacy.

2.4   SUMMARY

If there is anything this chapter should have demonstrated, it is that privacy is a complex concept—hiding many different meanings in many different situations in a simple seven-letter word. Without a clear understanding of what people expect from "having privacy," we cannot hope to create technology that will work accordingly, offering "privacy-aware" or "privacy-friendly" behavior. Law offers a good starting point for this exploration,[40] as it contains a society-sanctioned codification of privacy that goes beyond the views and opinions of individual scholars (such as the "GDPR" [European Parliament and Council, 2016]). However, understanding the *raison d'être* behind these laws—Why do they exist? What function do they serve in society?—offers further insights that help us shape the solution space for privacy-aware technology: privacy means empowerment, dignity, utility, a constraint of power [Lessig, 1999]; privacy functions as an emotional release, a human right, a staple of democracy, or as a driver for innovation [Clarke, 2006, Westin, 1967]. Last but not least, we discussed various conceptualizations of privacy from the literature (e.g., privacy constituents such as solitude, intimacy, reserve, anonymity, autonomy, control) that further completed the many possible uses for and benefits of privacy, and how the actions of others can affect this (e.g., Solove's privacy taxonomy).

---

[40]Note that our discussion on privacy law in this chapter is only cursory. An excellent overview of U.S. privacy legislation can be found, e.g., in Gormley [1992], Solove [2006], and Solove and Schwartz [2018]. For an international perspective, Bygrave [2014] offers a detailed discussion, while Greenleaf [2014] focuses explicitly on Asia, and Burkert [2000] on Europe. Voss [2017] and De Hert and Papakonstantinou [2016] specifically focus on the GDPR. Greenleaf [2017] offers a current overview of over 120 national and transnational privacy laws.

CHAPTER 3

Mobile and Pervasive Computing

Before delving further into an analysis of the privacy issues and implications of mobile and pervasive computing technologies, we first describe defining characteristics of both mobile computing and pervasive computing. While some of these characteristics, such as interconnectivity, are typical for computing systems in general, other characteristics, such as context awareness and implicit interaction, set mobile and emerging pervasive technologies somewhat apart from other information and communication technologies (ICTs). The interested reader may refer to Abowd and Mynatt [2000], Want [2010], and Ferscha [2012] for a larger historical perspective on the evolution and emergence of mobile and pervasive computing.

3.1 MOBILE COMPUTING CHARACTERISTICS

Mobile computing describes a paradigm shift from fixed, stationary computers and servers connected by wired networks to smaller, portable devices that users can take with them and interact with while on the go [Satyanarayanan, 1996]. Laptop computers, PDAs, and mobile phones emerged as the first mobile computing devices in the 1980s and 1990s. Since then mobile devices have evolved into smartphones and tablets, as well as more specialized devices, such as eBook readers, fitness trackers, smart watches, and smart glasses [Schmidt et al., 2012]. Four key aspects differentiate mobile computing from traditional computing: the (1) form factor and (2) computation and communication capabilities of mobile computing devices; (3) their ability to sense the environment, and (4) their software ecosystem.

3.1.1 NOVEL FORM FACTORS–MOBILITY AND DIVERSITY

In contrast to traditional computers, mobile devices are small and light enough to be portable and, therefore, live in close proximity to their users. Today, most of us carry a smartphone with us wherever we go. The relentless miniaturization of ICT components has not only led to portability, but also to diversity: many people today have multiple mobile devices that serve different purposes. For instance, a smartphone may be primarily used for social communication, quick look-ups, location-based services, and mobile entertainment; an ebook reader is used for reading longer texts; a tablet serves for entertainment and light work; a laptop supports more complex work tasks; a fitness tracker keeps track of one's step count and activity level; and a smartwatch provides us with just-in-time information and notifications.

Fitness trackers and smartwatches are a particular class of mobile devices that are worn on or at the body. Having long since been used only in labs and by a few pioneering individuals, such *wearable computers* [Billinghurst and Starner, 1999, Mann, 2013] recently saw huge commercial growth. Many manufacturers have released smartwatches that act as companions to smartphones, providing notifications and controls, e.g., about incoming calls or messages, and may also enables activity and fitness tracking with integrated sensors. Similarly, wrist-worn activity and health trackers are small sensor-equipped bracelets that can provide comprehensive information about the user's fitness and activity level. Slightly more futuristic yet increasingly available commercially are head-mounted displays such as Google Glass or Microsoft's HoloLens, which can augment a user's vision with information displays.

The sheer unlimited variety of shapes of current and future mobile computing devices means that users are not only able to carry them throughout the day, but increasingly also wear them at night. The diversity of purposes supported by such devices means that, in principle, some amount of computing and communication power is always within reach, even in situations where a general-purposes device may be impractical to use (e.g., while driving, sleeping, exercising). Last but not least, their portability and mobility also makes mobile devices more susceptible to loss, theft, and damage than fixed computing devices [Satyanarayanan, 1996].

3.1.2   POWER IN YOUR POCKET–COMPUTATION AND COMMUNICATION

As the "computing" in mobile computing suggests, mobile devices are nowadays equipped with substantial processing power. Consider for example, Apple's 2018 iPhone XS. The iPhone XS uses the A12 Bionic, a system on a chip (SoC) that contains six CPU cores, a dedicated 4-core graphics processor (GPU), a real-time machine learning engine with eight cores enabling immersive augmented reality experiences, and other components, while being energy-efficient [Apple, 2018]. The iPhone XS further has a range of sensors (e.g., barometer, three-axis gyro, accelerometer, proximity sensor, ambient light sensor) that enable continuous and energy-efficient processing of sensor data and motion-based activity detection.

Mobility necessitates untethered operation, and hence has driven wireless connectivity in recent years. Most mobile devices today support some sort of wireless communication: short-range (NFC, RFID, Bluetooth), mid-range (WiFi), and long-range (GSM, LTE) [Schiller, 2003]. Most communication is still using client-server patterns (e.g., Web and Cloud services), though ad-hoc peer-to-peer communication is increasingly common (e.g., a wearable device with a smartphone). Mobile devices and applications leverage these communication capabilities to facilitate ubiquitous access to information for users and synchronize information across devices and services. Mobile devices and applications must however account for variable and intermittent connectivity [Satyanarayanan, 1996].

However, not all mobile devices have the same processing and communication capabilities due to associated energy requirements and limited battery capacity in smaller devices especially. In contrast to traditional computers, the energy resources of mobile devices are finite [Satyanarayanan, 1996]. Smaller mobile devices, such as fitness trackers or smart watches, may be constrained to short-range or mid-range communication and have less powerful processors. Such devices typically off-load communication and processing tasks to other devices. They may connect to smartphones via Bluetooth or similar short-range protocols in order to let the more capable device perform processing on collected data and provide cellular communication to synchronize information with a cloud service, e.g., a fitness or quantified-self website. Smartphones may also hand-off tasks that require extensive data or processing to cloud services. For instance, most smartphones support powerful voice-based assistants, e.g., Siri on iOS or Google Assistant on Android smartphones, by forwarding recorded voice data to a cloud service, where powerful backends can use voice data from a large user base in order to improve voice recognition performance. Cyber foraging extends this approach by enabling mobile devices to opportunistically utilize available computing infrastructure in their environment [Flinn, 2012].

Advances in chip integration and low-power modes allow for the use of an incredible amount of processing power practically everywhere. The ability to connect to both other mobile devices and—via long-range communication—to any Internet service, allows mobile devices to not only outsource the heaviest computational tasks but also have almost limitless access to information. Such offloading and outsourcing inherently leads to much more widespread data sharing than ever before. For instance, voice assistants stream a user's voice queries from the user's phone to their company's servers to process and interpret the command [Nusca, 2011].

3.1.3   DATA RECORDING–SENSING AND CONTEXT-AWARENESS

Mobile computing was initially largely focused on the development of cheap and low power computing and wireless networking capabilities [Weiser, 1991, 1993]. To a large extent, such aspects have become commodities to be found in almost all mobile devices [Ebling and Baker, 2012, Weiser and Brown, 1997]. Today's innovation in mobile devices is often powered by numerous sensors that provide awareness of the device's context—and hence the user [Patel et al., 2006]. Today's smartphones can sense geographic location (using satellite-based, wifi-based, and cell tower-based positioning), orientation (compass, gyroscope), altitude (barometer), temperature (thermometer), and motion (accelerometer). Multiple integrated microphones and cameras can serve as audio and optical sensors, e.g., to detect ambient light conditions or noise levels, or to measure a user's physiological parameters like heart rate by placing a finger tip on the phone's camera lens [Pelegris et al., 2010, Scully et al., 2012].

This abundance of sensing has enabled a wide range of context-aware applications and services. Location-based services use the device's location to provide information about nearby points of interest (e.g., highly-rated restaurants, ATMs, public transportation), nearby contacts, or location-based reminders. Applications such as IFTTT (if this then that) [IFTTT, 2014] enable users to write their own triggers for certain contexts. Personalized mobile assistants such as Google Assistant [Google, 2014] or Cortana [Microsoft, 2014], as well as other context-aware third-party app launchers, provide location-specific and activity-targeted information based on the user's location, the user's activities (e.g., opened apps, sent text messages), calendar information, and other information sources (e.g., current traffic).

Context awareness is an integral part of the vision of ubiquitous and pervasive computing as it enables devices, applications, and services to adapt autonomously to the device's and user's context and activities [Schilit et al., 1994]. Today's mobile devices feature a wealth of sensing capabilities that already support a range of useful applications. The potential of such applications makes continuous data collection an increasingly attractive option for many (e.g., constant location tracking). While location information is still the primary context factor used in most context-aware mobile applications [Schmidt, 2012, Schmidt et al., 1999], other sensors are slowly starting to receive more attention, e.g., motion sensors to detect physical activity.

3.1.4   SOFTWARE ECOSYSTEMS–THE DEVICE AS A PLATFORM

A particular characteristic of today's smartphones and mobile devices is that they act as platforms for many different types of applications (apps). While this is nothing new in the context of PCs and laptops, mobile phones have traditionally been closed ecosystems that were fiercely guarded by carriers. The fact that today's smartphones and smartwatches can run software not only from the manufacturer (e.g., Apple or Google) or carrier (e.g., AT&T or Vodafone), but in principle from any third party, has greatly accelerated innovation in the mobile space. As of October 2018, 2.1 million apps are available for Android in the Google Play Store and 2 million for iOS in the Apple App Store [Statista, 2018]. Similar app ecosystems exist for wearables, such as smartwatches (e.g., Android Wear, Apple Watch, Pebble) and optical head-mounted displays (e.g., Google Glass).

Today's modern app ecosystems not only accelerate innovation, but also customization by users. Moreover, they greatly reduce the traditional power of carriers and manufacturers, leading to a democratization of the application space where in principle a single developer can easily reach millions of customers. However, as mobile apps have usually almost unfettered access to a device's communication, processing and sensing capabilities, as well as the user's information stored on the device, there is now a plethora of parties that are, at least in principle, able to closely monitor an individual's communication and information behavior. Consequently, permission management has since become an important aspect on mobile devices.

3.2   PERVASIVE AND UBIQUITOUS COMPUTING CHARACTERISTICS

In contrast to the mobile computing paradigm, which focuses on the shift from stationary computers to portable devices, the field of pervasive and ubiquitous computing[1] is driven by the quest for a more "natural" fit of computing to people's everyday lives. Originating from the vision of Marc Weiser in the late 1980s and early 1990s, pervasive computing aims to support users' activities and goals, rather than getting in their way by requiring users to focus on accurately operating and controlling a computer [Weiser, 1991, 1993]. Ultimately, pervasive computing complements and extends mobile computing characteristics with three novel dimensions: embeddedness, implicit interaction, and ubiquity.

3.2.1   EMBEDDEDNESS–INVISIBLE COMPUTING

Like mobile computing, pervasive computing strives to create ever smaller, yet powerful devices. However, pervasive computing aims to leverage the potential of miniaturization and commoditization of computing components much more substantially. The idea is to enrich physical artifacts—everyday objects such as cups or blankets—and environments—rooms, buildings, parks, plazas—by embedding sensing, processing, and communication capabilities into them. Instead of consciously interacting with a computer—even if it is a highly mobile one such as a modern smartphone—pervasive computing sees users interact with everyday items that are enriched with computing and communication power. This also has implications in terms of scale: instead of interacting with a single dedicated device, or even a small set of mobile devices, a pervasive computing environment may eventually contain hundreds of small-scale interconnected devices. Such devices can be part of the environment's infrastructure or be personal devices belonging to a specific user (e.g., embedded in clothing or personal artifacts) [Beckwith, 2003].

Local sensing and computing infrastructure can be combined with cloud services to facilitate synchronization and information exchange between different environments. Physical artifacts and the physical environment gain a virtual representation "in the cloud" that reflects the artifact's or environment's state and context. Such networks of interconnected physical artifacts and environments are also referred to as the "Internet of Things" (IoT) [Atzori et al., 2010]. In fact, IoT has become a catch-all term for advanced mobile and pervasive computing technologies.

Embedding pervasive computing technology into environments at a large scale enables comprehensive sensing of user behavior and personalized adaptation of such systems to the user's needs. Examples of such systems are smart and autonomous cars (or intelligent trans-

---

[1]While the terms "pervasive computing" and "ubiquitous computing" were initially characterized by nuanced differences [Want, 2010], nowadays both terms are used interchangeably and refer to the same research field and community. For instance, the two premier research conferences in this area, UbiComp and Pervasive—merged in 2013 to form the ACM International Joint Conference on Pervasive and Ubiquitous Computing (UbiComp); see http://www.ubicomp.org. In this book, we use the two terms interchangeably.

portation systems in general), smart homes (or smart buildings in general), and smart cities. IBM Corp. [2008] even coined the term "smarter planet." The "smartness" in these terms refers to the enrichment of living spaces with sensors and actuators that can sense—and ultimately predict—user behavior: to save energy, time, and user mind share.

A smart home is characteristically equipped with a set of sensors, activators, and computing facilities linking these components [Sadri, 2011]. Integrated sensors determine presence and activities of inhabitants, as well as measure physical characteristics, such as temperature or humidity. Activators, also called actuators, can change the state of the building according to sensed information, e.g., adjusting room temperature or switching on lights. These basic functions are often subsumed under the term *home automation* [Friedewald et al., 2005].

Ambient-assisted living (AAL) leverages smart home concepts to improve the quality of life for impaired and elderly individuals [Sadri, 2011] to facilitate prolonged autonomous, independent living in their own homes. AAL systems can support inhabitants in regularly taking medication, remind them of meal times, and prevent potentially dangerous situations, for example, by automatically turning off the stove after use [Moncrieff et al., 2008]. AAL systems can also monitor a user's health and vital signs and alert care takers about unusual conditions or accidents [Sadri, 2011].

Similar safety concerns drive much of the development behind smart cars [Jones, 2002, Silberg and Wallace, 2012]. Here, sensors are primarily meant to detect road conditions and share such information with close-by vehicles. Sharing is done both directly to other cars using short-range wireless communication, as well as by relying on road-side infrastructure (e.g., toll stations, bridges, street lamps) and cellular communication. Additionally, user interface elements such as heads-up displays and voice control focus on supporting drivers without distracting them from the surrounding traffic [Pfleging et al., 2012].

Pervasive computing components may not only be embedded into the user's environment but also into the user's clothing or jewelry. A smart shirt may measure the user's body temperature, heart rate, arousal, and other vital signs based on skin conductivity [Baig and Gholamhosseini, 2013, Lee and Chung, 2009] and share this information with a smartphone. A smart necklace may vibrate to notify the user of incoming messages or interesting deals offered in a nearby store. After smart glasses that can overlay the user's field of view with additional information ("Augmented Reality") [Starner, 2013], research is underway to embed this into smart contact lenses [Lingley et al., 2011]. Nano-scale computing and communications research aims to create devices small enough to be inserted into a human's blood stream, digestive systems, or organs to monitor and diagnose health issues without invasive surgery [Staples et al., 2006].

Note that the "invisibility" of such embedded computing does not necessarily have to involve physical size. Invisibility also has a cognitive dimension, where the interaction with pervasive computing components is integrated into the user's activity and goals in order to make the computer "disappear" [Weiser, 1991]. This disappearance hence is meant in a metaphorical sense, i.e., the user's expectations are met with minimal distraction from the envisioned task [Satyanarayanan, 2001]. Both cases—whether it is a physically invisible or "just" cognitively invisible computer—make it difficult for users to realize that they are interacting with a computer at all. While this is the intention of ubiquitous computing, it also means that maintaining individual awareness of data collection, processing, and dissemination activities becomes more difficult. Also, simple physical fixes for controlling data collection, e.g., sticking tape on a laptop's camera to prevent surreptitious recording, will not be feasible anymore.

3.2.2   IMPLICIT INTERACTION–UNDERSTANDING USER INTENT

Ubiquitous and pervasive computing constitute a major paradigm shift for human-computer interaction. Dedicated input and output components, such as mouse, keyboard, or a touch screen, are being gradually replaced with (or augmented by) more "natural" interaction modalities. Key activities in this space are *tangible user interfaces*, *natural interaction*, *multimodal interaction*, and *context awareness*.

Enriching everyday artifacts with sensing, processing, and communication capabilities allows us to create virtual representations of such artifacts and use them as "tangible" interfaces. Research on "tangible user interfaces (TUIs)" studies the association of physical artifacts with digital information, and how the manipulation of the physical object can be used to transform the associated information [Ishii and Ullmer, 1997]. Such direct interaction with physical artifacts can be leveraged to create more natural mappings between human-computer interaction and physical interaction [Abowd and Mynatt, 2000]. For example, instead of looking at a map to orient oneself, a tangible user interface would allow the user to look through a "lens" (e.g., a smartphone with an augmented reality map app) to see labels attached to landmarks they are seeing, or see navigation directions seemingly painted on the ground.

User attention is a limited resource [Roda, 2011], so if users are going to be surrounded by countless ubiquitous computing devices, these devices should not continuously compete for the user's attention. Instead, devices and applications must try to provide most of their output in the form of ambient, unobtrusive notifications in the user's periphery of attention [Weiser, 1991], yet easily be able to move to the user's center of attention if needed [Abowd and Mynatt, 2000, Weiser and Brown, 1997]. This implies that systems are becoming much more autonomous in their decisions, and that users intentionally will not be fully aware of the various system activities. "Natural interaction" aims for an adequate balance between autonomously acting devices and systems, and explicit user interaction and engagement.

"Multimodal interaction" seeks to enable users to engage with computing systems through multiple input/output channels [Dumas et al., 2009]. This allows users to communicate with a system in a way that is most conducive to their current activity. Common input modalities in such systems are speech and gesture recognition, as they free the user from having to focus on a specific input or output component in the environment. Visual, auditive, and tactile channels are used for output. In combination with pervasive projection of information into the environment, the user's whole environment can be turned into an immersive interaction environment. For example, Microsoft Research's RoomAlive project [Jones et al., 2014] uses multiple projectors and depth cameras to transform a user's living room into an immersive gaming experience in which walls and furniture are incorporated into the game visualization. The user can freely move in and interact with this game world through gestures and touching the physical and virtual objects.

Ultimately, pervasive computing seeks to provide systems and applications with "context awareness"—a nuanced understanding of the user's current situation, in order to provide more meaningful interaction experiences. Typically, multiple basic context features (e.g., location, physical movement, time of day, but also social context such as calendar entries or a user's social network) are combined to infer higher-level "situations" [Abowd and Mynatt, 2000]. Weiser stressed that ubicomp systems should be *smart* [Weiser, 1991], but they do not need to be actually intelligent. A "smart" coffee maker might only need a few context features to understand if it should start brewing a fresh cup of coffee (e.g., time of day, day of the week, and the last 15 min of user movement). Typical context features include location, time, the user's activity, present persons or devices, and other information available about the user, such as their schedule. However, a high level of context adaptivity also requires a deeper understanding of the user's activity and personal experiences [Dourish, 2004]. In order to form such an understanding, ubiquitous computing systems can try to *adapt* their behavior over time: initially, users are involved in individual decisions yet gradually the system moves toward automated decision making [Bardram and Friday, 2009]. Thus, ubicomp systems would not only adapt to context changes, but also adapt their behavior to individual users.

The ability of ubicomp systems to be "smart" relies not only on the availability of context information, but also on the accuracy of this data [Bardram and Friday, 2009]. Ideally, systems should autonomously learn to cope with previously unknown situations, users, and other entities at runtime [Caceres and Friday, 2012]. *Ambient intelligence (AmI)* and *intelligent environments* are terms that are supposed to highlight this evolution from simple smartness to a deeper understanding of user context. AmI also underlines a trend toward more proactive systems consisting of agents that act autonomously on the user's behalf [Caceres and Friday, 2012]. Such proactive systems try to look ahead for the user by combining disparate knowledge from different system layers [Satyanarayanan, 2001]. Consider, for example, a personal digital assistant that automatically reschedules appointments when the user is stuck in traffic, or the notorious smart fridge that keeps track of available food and automatically reorders fresh produce as needed [Langheinrich, 2009]—potentially taking into account the users eating habits and preferences, as well as dietary and health considerations.

The combination of novel input and output capabilities, together with a system's awareness of user context, has enabled a novel interaction paradigm: *implicit interaction* [Schmidt, 2000]. Sensor-driven awareness of the user's context and behavior is interpreted as input to provide situation-specific support and adaptation [Schmidt, 2000]. Research on affective computing further aims to recognize a user's emotions in order to adapt systems with respect to the user's mood [Picard, 2003]. Depending on the application, the user's behavior may determine the system's reactions or the system may proactively modify the user's environment. Thus, instead of acting *with* a computer, the user acts *within* a ubicomp system and is surrounded by it. Interaction with such a system becomes continuous, i.e., interaction has no defined beginning or ending anymore, and may be interrupted at any time [Abowd and Mynatt, 2000]. This continuity and ubiquity of ubicomp applications facilitates the support of everyday tasks [Weiser and Brown, 1997]—tasks and problems that relate to and occur in the daily routines of users.

Two main implications stem from such interaction models. First, data collection becomes paramount for "smart" or even "intelligent" systems. This not only drives an effort to deliver ever greater accuracy in sensing (e.g., high quality audio and video recording), but also a continous urge to include more sensing modalities in order to not "miss" a crucial piece of context. Second, understanding the complexities of everday life seems to require collecting an ever-increasing share of a user's life, i.e., both spatially and temporally.

3.2.3   UBIQUITY–FROM SOCIAL TO SOCIETAL SCALE

The terms "ubiquitous" and "pervasive" already express the widespread presence that sensing and computing devices should have in a future with ubiquitous computing. Such a presence will allow for a new level of "social" and sociotechnical systems that offer highly customized services, based on intimate observations of individuals. At the same time, the ubiquitous availability of sensing and computing will also prompt applications at a "societal" scale, i.e., large-scale deployments that will affect cities, regions, and countries.

A prime example of this is the "smart grid," i.e., optimizing and coordinating energy consumption across multiple households and neighborhoods to stabilize the power grid against peaks. It envisions that individual smart appliances, such as dishwashers or washing machines, coordinate with energy providers to shift their workloads to times when surplus energy is available. This not only saves money for the consumer but also eliminates consumption peaks as energy providers can use monetary incentives to much excert more fine-grained control over demand. Another example is "smart transport," where cars periodically broadcast their location, speed, and heading to enable collaborative collision avoidance between individual vehicles [Kargl, 2008], as well as provide a detailed overview of the traffic situation to improve traffic prediction and control. When deployed at large-scale, personal devices such as smartphones or even wearable computers can learn not only about their owners but also—collaboratively—about the behavior of larger groups and societies [Lukowicz et al., 2012]. For example, the activity recognition running on a smartphone (in order to understand a user's current activity) can be made remotely available in order to allow for "opportunistic sensing"—the dynamic brokering of sensor information in order to collect information about a certain area, e.g., the level of crowdedness at a shopping mall, or the air quality across a city [Das et al., 2010, Ganti et al., 2011, Lane et al.]. Such large-scale, cooperative sensing with mobile devices can provide a "sociotechnical fabric" [Ferscha, 2012], which can provide a powerful analysis of social interactions, generate novel models of human behavior and social dynamics, and spur the development of socially-tuned recognition algorithms [Lukowicz et al., 2012].

A major challenge in achieving effects on a societal scale is the connection of many heterogeneous context-aware entities into large-scale ensembles of digital artifacts [Ferscha, 2012]. In addition to establishing inter-connectivity between heterogeneous systems and devices, self-organization and cooperation of those entities are paramount challenges in large scale complex dynamic systems [Th. Sc. Community, 2011]. Ensembles of devices with emergent and evolutionary capabilities can form societies of artifacts in order to support specific activities or user goals. Novel engineering and programming concepts are required to leverage the potential of these developments [Th. Sc. Community, 2011].

Embeddedness and continuous interaction will soon allow for long-term interactions with computers that will be much more intimate than any other data collection to date. However, with the realization of true inter-connectivity between heterogeneous systems, novel forms of social monitoring and control will not only cover the individual, but also extend to social groups and entire societies.

3.3 SUMMARY

Modern computing systems in general are characterized by a high degree of interconnectivity. However, mobile and pervasive computing are significantly different from "traditional" computers (e.g., a laptop or a desktop computer) due to seven reasons.

1. *Novel form factors:* Computers now not only come as powerful smartphones but are also embedded in clothing and toys, making it possible to have them with us almost 24 h a day.

2. *Miniaturized computation and communication:* Today's miniaturized computing resources allow us to run sophisticated machine learning applications in real time, or wirelessly transfer large amounts of data at gigabit speeds.

3. *Always-on sensing:* Modern sensors not only use fewer power than ever before, but also include sophisticated digital signal processors that provide application developers with usable high-level context information (e.g., indoor and outdoor location, physical activity).

4. *Software ecosystems:* App stores have revolutionized the way we distribute and consume software. Never before was it easier to bring new software to millions of users, yet users now need to be better trained to understand the implications of installing untrusted programs.

5. *Invisibly embedded:* The low costs of computing, communication, and sensing systems has made it possible to make rooms, buildings, and even entire cities "smart." As this ideally happens without distractions (e.g., blinking lights), it will become increasingly more difficult to tell augmented (i.e., computerized) from un-augmented spaces.

6. *Implicit interaction:* The ubiquity of computers has made it possible to create "invisible" assistants that observe our activities and proactively provide the right service at the right time. Obviously, this requires detailed and comprehensive observations.

7. *Ubiquitous coverage:* Embedding computing from small-scale (e.g., in our blood stream) to large scale (e.g., across an entire metropolitan area) significantly increases both vertical and horizontal coverage (i.e., across time and space) of our lives.

While today's interconnectivity certainly forms the starting point for most of today's privacy issues, the characteristics discussed in this chapter are of particular interest when looking at the privacy implications of mobile and pervasive computing. The next chapter will discuss these implications in detail.

CHAPTER 4

Privacy Implications of Mobile and Pervasive Computing

In his seminal 1991 Scientific American article, Mark Weiser already cautioned that *"hundreds of computers in every room, all capable of sensing people near them and linked by high-speed networks, have the potential to make totalitarianism up to now seem like sheerest anarchy"* [Weiser, 1991]. Chapter 3 presented some of the reasons for this: mobile systems come in highly portable form factors that make it easy to always carry them with us; their powerful communication capabilities encourage data offloading to the cloud; the potential of context awareness and novel low-power sensors make continuous data collection the default; and thriving app ecosystems challenge traditional trust relationships. The vision of pervasive systems furthermore makes it difficult to tell when one is detected and potentially recorded by invisible devices; its focus on understanding user intent drives ever expanding data collection; and the ubiquity of smart devices and environments offers the tantalizing promise of better "managing" and "optimizing" society.

Data collection and processing are core aspects of mobile and pervasive systems and they create a dual-use dilemma. Consider the example of the smart fridge that automatically re-orders food from the grocery store. This functionality clearly benefits the user, as it obviates the need for last-minute trips to the grocery store. However, at the same time, the consumption patterns and eating habits implied in the fridge's store orders *could* be used by the store for targeted advertising, or even to create personalized adjustments of prices. Assuming that the user sees this more as "manipulation" than a useful service, this would clearly be a drawback [Langheinrich, 2009]. Even more critical, such detailed grocery lists are likely sufficient to infer a person's (or a whole family's) health risks, something that an insurance company might be interested in buying from the store as part of a background check before offering a new insurance contract. Obviously, this could also be viewed as beneficial: a consumer with a smart fridge who keeps a healthy diet might pay less for health insurance if the individual would explicitly allow such data sharing; maybe also in exchange for additional benefits, such as personalized health tips and recipes.

Weiser [1991] noted that *"fortunately, cryptographic techniques already exist to secure messages from one ubiquitous computer to another and to safeguard private information stored in networked systems."* However, the mere existence of cryptographic techniques is obviously insufficient to address the problem of privacy. Take email for example: more than 25 years later we are still unable to effectively safeguard the transmission of personal email on any useful scale. Safeguarding complex systems takes far more than cryptography [Anderson, 2008]. In addition, as we discussed in Chapter 2, simply *"ensur[ing] that private data does not become public"* [Weiser, 1991] is not sufficient to provide privacy [Langheinrich, 2009].

In this section, we explore the specific privacy implications of mobile and pervasive computing technology, in order to determine the challenges that need to be tackled if we want to provide privacy-friendly technology, systems and applications. Much of what we discuss here of course also applies to computers, servers and cloud infrastructures. In fact, Abowd [2012] argues that ubiquitous computing has become almost indistinguishable from general computing, as most recent computing advances could also be considered ubicomp advances and vice versa. We thus start each section by discussing the general implications of computers with respect to privacy, before moving on to focus on mobile and pervasive applications in particular.

Privacy implications of large-scale data collections have been recognized as early as the 1970s (see Section 2.1.1). Paul Sieghart, one of the authors of the influential UK White Paper on Computers and Privacy in 1975 (see Douglas [1976]), described the effects of the computerization of daily life—the "information society"—as follows.

> More transactions will tend to be recorded; the records will tend to be kept longer; information will tend to be given to more people; more data will tend to be transmitted over public communication channels; fewer people will know what is happening to the data; the data will tend to be more easily accessible; and data can be manipulated, combined, correlated, associated and analyzed to yield information which could not have been obtained without the use of computers [Sieghart, 1976].

Computers are responsible for three core developments that greatly shape our information society: the "digitization" of everyday life; the automation of capturing real-world processes; and profiling.

4.1   DATA SHADOWS–THE DIGITIZATION OF DAILY LIFE

The first core driver of how computers affect privacy is their ability to "digitize" our daily lives— to map the complexity of the real-world to a set of bits. The phenomenon of digitization of our lives began in the 1960s and 1970s, when the first databases allowed governments to take stock of their citizens—not only in terms of population numbers, but also their demographics and how they live—through large-scale censuses. However, censuses only captured a single moment in one's life and relied on self-reported data, making the quality of the mapping relatively coarse. Today's digital traces are much more comprehensive.

In the following (and in subsequent sections in this chapter), we will first briefly describe the technology behind each of the core developments (digitization, automated capture, and profiling), before discussing their privacy implications—both for computers in general, and in particular given the capabilities of mobile and pervasive computing.

4.1.1 TECHNOLOGICAL DEVELOPMENT

The early censuses of the 1970s were the first attempt to build a digital representation of a country's citizens. Today, the often-quoted "transparent citizen" has become a reality, as we will illustrate in five sample domains: payment systems, interpersonal communication, media consumption, physical movement (transport), and physical activity.

One of the earliest commercial drivers of digitization were cashless payments in the form of credit and debit cards. Card-based payment systems increased digitization considerably, both in terms of detail and reach. At first, mostly a tool to arrange international travel (e.g., hotel and rental car booking), card-based payment systems have since expanded to also cover many of our everyday purchases. Innovations such as NFC-based contactless payments—either with a payment card or a compatible smartphone—seek to lower the barrier of use for such systems in order to encourage cashless payments even for small everyday purchases (e.g., a cup of coffee or a bus ticket). E-Commerce was an early driver of this process, as cashless payments are often the only way to transact business online. While consumers value the convenience of cashless payment (less small change to carry around; no need to have enough cash with you; better protection from theft), both industry and government also significantly profit from the traceability of transactions: cashless purchases enhance the creation of consumer profiles and thus improve marketing, while fewer cash transactions mean that less money can potentially be hidden from the tax office.

The second wave of digitization came in the form of email and text messaging (e.g., through pagers and texting). Written digital communication has moved much of our daily interactions—some of which we might have had face-to-face or over the (analog) phone—into the digital realm. Digital communication offers countless benefits: the ability to chat from a multitude of devices, often free of charge, including the exchange of images, files, and video; the ability to exchange arbitrary amount of text, as well as documents and pictures in high quality with practically anybody in the world; the ability to communicate quickly yet asynchronously (i.e., no need for the conversation partner to be available at the same moment). Later on, the ability to transport voice over the Internet (VoIP) allowed not only telecommunication providers to lower their infrastructure cost, but also allowed end-users to both receive and field calls on their mobile, their computer, or a (VoIP-enabled) desktop phone, depending on their current location and preferences. Finally, video telephony—the vision from the 1960s that just never seemed to catch on—took off when software such as Skype transformed any computer with a cheap webcam into a telepresence device. Video conferencing, often from smartphones, is know common for both business and personal interactions. Digital communication has truly transformed our ability to keep in touch and work together across time and space. Yet digital communication inherently allows for the capture of detailed "connection metadata," i.e., who communicates with whom, when, and for how long, and, depending on the communication means, also its contents. Even "normal" landline telephony is nowadays mostly implemented digitally (i.e., as VoIP), as, of course, is mobile telephony, which means that meticulous call records can be easily maintained by telecommunications companies and government agencies—including foreign and domestic intelligence agencies, as evidenced by the Snowden revelations on the large-scale data collection by the NSA, GCHQ, and other intelligence services [Landau, 2013, 2014].

The development of digital formats for media—initially ebooks and subsequently music and video—represents a third wave of digitization. Today, music and videos are more and more delivered and consumed online via streaming services, while ebooks and their corresponding reading devices are replacing traditional books. Music streaming has started to fundamentally change the media consumption behavior of a generation, as more and more young people do not buy individual albums or even songs anymore (and certainly not on a physical medium such as a CD!) but simply pay a monthly fee for accessing a more or less unlimited amount of music. The video industry similarly uses video streaming to both simplify the distribution process (no lengthy downloads of huge video files) and to combat piracy (as no digital file is ever available to customers for illegal sharing). Increasingly, even traditional TV content is—now in a fully digital format—being provided on-demand (e.g., Netflix) or consumed through cloud-based streaming services that allow consumers to pause live content or re-watch missed shows. Last but not least, ebooks and other digital content are not only often cheaper than printed books and physical media (certainly for the publisher as no actual printing and shipping takes place, but also for the consumer) but can also be created on-demand and instantaneously (hence no warehouses needed for storage) and can be carried around by the thousands on a small memory card (thus allowing consumers to take an entire library to the beach).

The flip-side of streamed media is the ability for publishers to track listening and watching habits. While ebooks in principle do not suffer from this problem (they are small enough to be downloaded in their entirety), ebook readers offer consumers to keep track of their books and the current page through cloud services that sync reading state across ebook readers and reading apps on mobile devices. This information could be easily shared with publishers to provide insight not only on how many books are sold but also how long it takes people to read a book or where they stop reading.

Many consumers also do not realize that acquiring electronic media often does not constitute ownership, as it was the case with regular books or records or DVDs and Bluray discs. Instead, consumers typically license (i.e., "rent") a certain item. This means that publishers cannot only at any point remove access (e.g., delete the ebook from the reading device [Stone, 2009]), but also that re-sale or gifting such media is typically prohibited [Bogle, 2014].

A fourth example of everyday digitization can be found in transport. Both airline travel and railway companies pushed the concept of digital "print-at-home" tickets, while public transport companies are increasingly supporting "on-the-go" payment schemes via NFC-enabled cards and smartphone apps. Such digital records thus follow our international, national, and even local traffic in ever increasing detail. Many car rental companies today track the position of their vehicles, not only to prevent theft, but also to enforce adherence to road safety [Mc-

Garvey, 2015]. Several car manufacturers offer roadside assistance programs that continuously track the position of the car and automatically alert emergency services in case of an accident (e.g., if the airbags are triggered). Moreover, modern cars are increasingly driven "by wire," i.e., steering and pedals are analog devices whose input is digitized and used to control motors that adjust the angle of the wheels, the pressure of breaks, and the amount of fuel injected into the motor. Countless sensors collect additional information on road conditions (e.g., temperature and slipperiness), allowing for the detailed data capture of an entire road trip [Musk, 2013].

Note that location capture is not restricted to transportation: many door locks—not only in offices and hotels, but also private residences—do not rely on physical keys anymore but use remotely-programmable chip cards or mobile apps. Every time someone opens them, another event is added to the data log. Internet-connected locks and security cameras provide detailed audit logs of who is coming and going in our homes [Ur et al., 2014]. In fact, today's home automation solutions already explicitly target the capture of home presence patterns in order to optimize heating systems, while electronic "smart meters" are being deployed to provide utility providers with instant information about energy and water use. If fine-grained enough, such energy data cannot only detect use of individual appliances but even allow inference of the TV program one is watching [Greveler et al., 2012]. Such detailed records of our activities can also be inferred from the data we actively publish about ourselves: many people today continuously update their social media "status" to share their activities with others on social networking sites, such as Facebook, Instagram or Twitter. Those status updates may not only include a textual description but also location information, pictures, and explicit co-location information of others.

In short, most of our life today has a digital "data shadow" that represents an electronic record of our "analog" reality. Such data shadows are a core enabler of modern life, as they allow us to efficiently deal with much of its inherent complexity. They enable our highly mobile lifestyles, allowing us to communicate on-the-go (instead of having to wait for physical letters to arrive or finding a fixed-line telephone) and to organize our business and our private lives wherever we are (using mobile apps on our phones or tablets for everything from company work to private banking or home automation). The ability to access these data shadows from almost anywhere also supports our interactions with others, e.g., when we use credit cards to pay in a foreign country. The concept of "anticipatory computing" [The Economist, 2014] uses our data shadows to anticipate our future plans and behavior in order to provide us with information right when we need it, e.g., reminding us when to leave for an appointment based on our current location, the appointment's location, and the current traffic conditions between them [Google, 2014].

4.1.2 PRIVACY IMPLICATIONS

While digitization is a key enabler for our highly efficient lifestyles, it does come with significant privacy implications that are only exacerbated by mobile and pervasive computing, for two main reasons.

First, mobile and pervasive computing poses a shift in what kind of information is collected about users and at what scale [Langheinrich, 2009] compared to previous databases or even data collection on the Internet. The creation of such data shadows relies on the collection of "real-world" data about us and the digital storage of this data. Both mobile devices and pervasive computing systems greatly simplify the continuous collection of such day-to-day information about an individual. Rather than requiring users to explicitly provide information, sensors in mobile and pervasive environments enable the invisible collection of information [De Hert et al., 2009], extending to location, health, and behavioral information about specific individuals. Sensing here does not necessarily entail a physical sensing device—one of the most revealing "sensors" is a piece of plastic: the credit card. In today's commercial environment, the increasingly ubiquitous credit cards provides detailed movement and consumption data from which further inferences, such as our behavior and activities, can be drawn [Ackermann and Mainwaring, 2005]. The potential for extensive realtime monitoring and permanent recording of minutiae of a user's everyday life would lead to *"surveillance of the ordinary"* [Langheinrich, 2009] on a pervasive scale [Cas, 2005, De Hert et al., 2009]. This issue is exacerbated by the fact that associated information flows and uses of data are obscured from users [Lederer et al., 2005], due to the embedded nature of many mobile and pervasive computing systems.

Second, today's cheap and advanced storage capacities [Waldo et al., 2007] make it highly likely that the collected information is stored permanently. Waldo et al. [2007] note that developing procedures for limiting data retention and deleting data is often considered more expensive by companies than just keeping the data. As a result, previously transient information becomes permanent, which crosses the ephemeral and transitory borders defined by Marx [2001] (see Section 2.3.3) [Bohn et al., 2005, Langheinrich, 2002b]. Once collected and never deleted, data is thus much more likely to be used for emerging purposes that were not considered or available at the time of collection. For example, when an earthquake hit California in August 2014, fitness wearables manufacturer Jawbone analyzed sleep tracking data collected by its users' fitness trackers to study how the earthquake affected the sleep patterns of people living in different parts of California [Mandel, 2014]. Google uses aggregated information about certain health-related search terms to estimate flu activity and detect flu epidemics in the United States [Ginsberg et al., 2009]. While in both cases analysis results are based on aggregated data, it is conceivable that the same data could be misused by healthcare providers or employers to determine health risks of individuals, which could then be reflected in increased insurance costs, or decreased chance of being considered for a promotion.

4.2   FROM ATOMS TO BITS–AUTOMATED REAL-WORLD DATA CAPTURE

The second core driver of how computers affect privacy is our increasing ability to seamlessly capture real-world events. Traditionally, digitizing data meant that information had to be entered manually into a computer system. With the help of additional computing power, such "media breaks"[1] can be greatly reduced. An example of a media break would be when goods arriving at a warehouse are not properly added to the inventory database [Fleisch et al., 2003]. Computing has long been a key component for reducing media breaks along the supply chain. For example, instead of entering prices manually into a cash register, a barcode allows a properly equipped cash register to not only avoid errors in producing the total price, but also to keep track of the product being sold, not just its price. Mobile and pervasive computing will greatly improve this process and allow information processes to further close the gap between the real world and the virtual world. A truck's on-board tracking unit (OBU) can provide logistics companies with instant fleet management, allowing them to know at any point in time where their trucks are. Similar units mounted to shipping containers have revolutionized international shipping, making it much easier to track individual shipments across the globe. On a smaller scale, delivery agents scan barcodes with their mobile devices when they process or deliver a package, allowing a parcel service to instantaneously update the shipping status of each parcel.

4.2.1   TECHNOLOGICAL DEVELOPMENT

The ability of mobile and pervasive computing to eliminate "media breaks" also applies across other areas of life, beyond the industrial supply chain. In fact, the massive *digitization* of our lives described in the previous section is only possible due to our ability to use mobile and pervasive computing to track and capture real-world processes in almost real time. Smartphones in our pockets capture our position instantaneously and continuously, making it possible to record comprehensive movement logs over days and months. Moreover, collected sensor data is also easily shared between applications and even across devices. A mobile browser can easily query a mobile device for its location in order to provide localized search results using HTML5's geolocation API [Mozilla Foundation, 2018]. An application on a mobile phone can query a body-worn health sensor for physiological data (e.g., heart rate) and use it to provide exercise advice and dietary suggestions [Katz, 2015]. Centralized profiles such as one's Google, Apple, Amazon, or Microsoft account allow one to continue an activity, e.g., a browsing session, started on a desktop computer later on with a mobile device or continue a chat on a mobile device on one's laptop.[2] Sensors in vehicles, ships, and airplanes allow for the servicing of engines and other parts even before a potential failure occurs, simply by continuously logging and transmitting key operational parameters to a service center [Tel, 2010]. Similar technology has long been in use in vending machines to alert the operator of potential out-of-stock situations before they occur,[3] while early prototypes already exist that try to offer a similar functionality for supermarket shelves [Koesters, 2018].

---

[1]"Media break" is a term from (German) business informatics that describes a missing link in the information flow of a business.

[2]For example, Apple Continuity (`https://www.apple.com/macos/continuity/`) enables activity transitions among different Apple devices.

[3]For instance, NetVends (`http://www.netvends.com`) offers such remote vending solutions.

Modern sensor technology is key for eliminating media breaks. Today's sensors use less and less power, making it possible to both embed them into ever smaller packages (as they do not need a large battery) and to run them for longer periods of time. For instance, positioning information used to be available only through the use of power-hungry GPS sensors, which meant that consumers had to choose between battery life or detailed localization. Today, those GPS sensors have not only become much more power efficient, but they are also used together with accelerometer sensors to better understand when the device is actually moving and hence position information needs to be updated. Power-efficient WiFi chipsets complement GPS-based location information—in particular indoors—by using WiFi fingerprinting technology [Husen and Lee, 2014]. Some sensors can also harvest the required power from the measurement process itself, making it possible to forego a battery completely.[4] Alternatively, infrastructure-based sensors "piggyback" onto the power grid or plumbing of a house and infer occupancy information or individual device use simply by observing consumption patterns [Cohn et al., 2010, Froehlich et al., 2009, Gupta et al., 2010, Patel et al., 2007, 2008].

4.2.2   PRIVACY IMPLICATIONS

Continuous sensing significantly changes not only the accuracy of data collection, but also the manner in which such data collection is taking place. Instead of closing a shop and manually taking an inventory of what is on its shelves, a sensor-equipped smart shelf "knows" at any point in time what products are on its shelves. While this is of course greatly reducing costs (e.g., by reducing the above-mentioned "media breaks") it also has implications for human perception of such collections. The more these "points of capture" move into the background, into the fabric of the infrastructure, the less awareness can there be for such processes. A "smart home" may need no cameras to know what its inhabitants are doing at any point in time, as the electric wires embedded in its walls are able to pinpoint the exact location of each person within a few meters [Adib et al., 2015]. As mentioned in the previous section, a "smart meter" enables the utility company to remotely gather information a household's power consumption, usually in real-time, with the potential to infer in real-time household presence, appliance use [Gupta et al., 2010], and what TV program is being watched [Greveler et al., 2012]. A "smart car" may continuously relay its current location, speed, number of passengers, and even the radio station currently tuned into, to a central traffic system or the car's manufacturer, without any speed cameras, surveillance cameras, or toll checkpoints around. Sensor-based systems have made data collection so easy that it has become the default, not the exception.

This always-on sensing (and collecting) has significant security implications. In a 2015 report [Federal Trade Commission, 2015], the US Federal Trade Commission estimated that today's nascent "Internet of Things" already connects over 25 billion devices, and points out that companies with a large store of consumer data will become "a more enticing target for data

---

[4]See, for example, products by EnOcean, https://www.enocean.com/.

thieves or hackers." A second implication of such seamless data collections is that it exacerbates what Solove [2013] calls the "consent dilemma."

Consent has been one of the cornerstones of modern privacy legislation (see Section 2.1.1), stipulating that many data collections are legal if the data subject has (explicitly or implicitly)[5] given their consent. In fact, the latest EU privacy law, the GDPR (see Section 2.1.2), has significantly increased the requirements for data collectors to obtain consent (Article 4.11, GDPR).[6] Solove points out two key problems with consent. First, Solove points out *cognitive problems* that severely "undermine …individuals' ability to make informed, rational choices about the costs and benefits of consenting to the collection, use, and disclosure of their personal data." Second, Solove identifies what he calls *structural problems* of asking individuals for consent: first, there are "too many entities collecting and using personal data to make it feasible for people to manage their privacy separately with each entity;" and second, many privacy harms are "the result of an aggregation of pieces of data over a period of time by different entities. It is virtually impossible for people to weigh the costs and benefits of revealing information …" In a sensor-based environment, giving "unambiguous" consent may be impossible for two reasons: first, there may not be any explicit dialog between the system and the user, e.g., in a smart office building that automatically tracks all visitors; and second, the sheer number of service may overwhelm the cognitive capacity of users, who, according to Solove [2013], already are barely able to keep track of these data collections within the Web context.

4.3   PROFILING–PREDICTING BEHAVIOR

Profiling is a practice that in principle is as old as human relationships [Solove, 2008]. In their "Advice paper on essential elements of a definition and a provision on profiling within the EU General Data Protection Regulation," the Article 29 Data Protection Working Party [2013][7] defines profiling as follows.

> "Profiling" means any form of automated processing of personal data, intended to analyze or predict the personality or certain personal aspects relating to a natural person, in particular the analysis and prediction of the person's health, economic situation, performance at work, personal preferences or interests, reliability or behavior, location or movements.

Profiling is one of the most critical aspects of any privacy law today, as its potential for both benefiting users—in the form of better services—*and* harming users—by increasing, e.g., their vulnerability to manipulation—is tremendeous.

---

[5]An example of implicit or implied consent is an "opt-out" system that pre-ticks a "I consent" box but allows users to deselect it. Explicit or informed consent would require users to actively opt-in, i.e., manually tick said "I consent" dialog box.

[6]It now requires an "unambigious" indication by "a statement or by a clear affermative action"—the previous Privacy Directive 95/46/EC only required "informed" consent.

[7]The Article 29 Working Party was a European data protection advisory body consisting of representatives from the data protection authorities of each EU member state and the European data protection supervisor. With the GDPR, the Article 29 Working Party has been reconstituted as the European Data Protection Board.

4.3.1 TECHNOLOGICAL DEVELOPMENT

In the past, small store owners would get to know their customers over time and begin to antici-pate their needs, e.g., by pre-ordering the customer's favorite items. Computers allow this predic-tion to become automated and externalized, allowing the operator of such a profiling operation to predict a person's desires, fears, or actions without having even met the person. Today, informa-tion collected online, e.g., on e-commerce sites, forums, and search sites, is routinely integrated with databases containing detailed records about our offline lives, e.g., household income, pro-fession, marital status, credit scores, in order to, e.g., assess individual affluence, interests, and creditworthiness—all in real-time, as we are "surfing the Web" or using our smartphones. Such "online behavioral advertising (OBA)" tracks individuals across websites and online services to infer their interests and estimate their propensity to purchase specific products and, thus, their susceptibility to targeted advertising [Rao et al., 2014]. Online behavioral information may fur-ther be enriched by an individual's purchase history from multiple websites, as well as offline purchases. For instance, in the United States data brokers legally obtain information about a person's prescription drug purchases and usage related information, compile it into prescription drug reports, which are then bought by insurance companies to estimate health risks of appli-cants and to decide whether they will be insured [Privacy Rights Clearinghouse, 2012]. There is an active market for personal information collected online and offline [Schwartz, 2004]. A par-ticularly morbid illustration of what information is stored in such profiles came to light in 2014, when Mike Seay, whose teenage daughter had died nine months earlier, received a junk-mail letter from OfficeMax with a second address line reading "Daughter Killed in Car Crash" [Hill, 2014].

The information collected by mobile and pervasive computing systems makes even more fine-grained information available for aggregation, but also provides additional data points to attach external information to. Fine-grained sensor, location, and activity data can be corre-lated with public or personal events, venue information, and social media activity. The result will be comprehensive and holistic profiles that map out a person's life. Today's online social networks, with their wealth of pictures posted from a huge variety of social situations (holidays, meetings, parties, sports, events), already allow for unprecedented aggregation, consolidation, and de-anonymization. Advanced machine learning methods allow us to classify human inter-ests and personalities from seemingly unrelated information, such as call logs or texting behav-ior [Chittaranjan et al., 2011]. For example, Acquisti et al. [2014] were able to infer both the interests and the social security number[8] of total strangers only from their picture—by using face recognition, data mining algorithms, and statistical re-identification techniques.

---

[8]In the U.S., the social security number is a de-facto universal identifier that is used as an authentication token in many situations. Knowing a person's name and social security number is usually enough to impersonate that person in a range of situations (e.g., opening accounts, obtaining a credit card, etc.) [Berghel, 2000].

4.3.2   PRIVACY IMPLICATIONS

Information aggregation and comprehensive profiling have multiple potential privacy implications. An obvious issue is that profiles may contain incorrect information. Factual information may have been inaccurately captured, associated with the wrong person due to identical or similar names, or may have been placed out of context as part of aggregation. Inferences made from collected data may be incorrect and potentially misrepresent the individual (see, e.g., Charette [2018]). Inaccurate information in an individual's profile may just be "a nuisance," such as being shown improperly targeted ads, but consequences can also be dire, such as having to pay a higher premium for health insurance, a higher interest rate for a loan, being denied insurance, or being added to a no-fly list. Some options exist to correct inaccurate information. Credit bureaus provide mechanisms to access and correct credit reports; online data brokers often provide access to one's profile and may allow for corrections [Rao et al., 2014]. However, correcting inaccurate information can still be difficult or even impossible because it is difficult for an individual to determine the original source of some misinformation, especially if it is used and disseminated by multiple data brokers.

Even if information aggregated about an individual is correct, privacy issues arise. Based on a profile's information individuals may be discriminated against in obvious as well as less perceptible ways. Price discrimination is a typical example. For instance, the online travel agency Orbitz was found to display more expensive hotel and travel options to Apple users than Windows users based on the transferred browser information [Mattioli, 2012]. Acquisti and Fong [2014] studied hiring discrimination in connection with candidate information available on online social networks. They find that online disclosures of personal traits, such as being Christian or Muslim, significantly impacted wether a person was invited for a job interview. It is imaginable that inferences about an individual's health based on prescription drug use, shopping history (e.g., weight loss pills or depression self-help books) or other indicators may also lead to discrimination. A key approach in addressing the risks stemming from profiling is thus also the need to disclose information about the algorithm(s) used to rank or classify an individual (see Section 5.6).

Large-scale collection and aggregation of information can also lead to inadvertent disclosure of some information about individuals that they would have preferred to keep private. For instance, Jernigan and Mistree [2009] used social network analysis to predict with high accuracy a person's sexual orientation based on their friends' sexual orientation disclosed on online social networks. Information shared in mobile messaging apps about when a user is active or "available to chat" is sufficient to infer a user's sleep times, chat partners, and activities [Buchenscheit et al., 2014]. Mobile devices and laptop computers continuously send service announcements in order to enable interconnectivity and seamless interaction between devices, but may also leak a user's presence and identity when devices are connected to open wireless networks [Könings et al., 2013].

4.4 SUMMARY

Three fundamental trends in mobile and pervasive computing have a significant impact on our privacy: the digitization of our everyday life; the continuous data capture with the help of sensors; and the construction of detailed profiles. While none of these trends are new, mobile and pervasive computing exacerbate these issues greatly. As a consequence, we have an ever increasing amount of information captured about us, often well beyond what is directly needed. The ability to use advanced sensing to collect and record minute details about our lives forms the basis for detailed personal profiles, which, with the help of data mining techniques and machine learning, provide seemingly deep insights into one's personality and psyche. At the same time, this information allows for unprecedented levels of personalized systems and services, allowing us to manage an ever-increasing amount of information at an ever-increasing pace. Left unchecked, however, these powerful services may make us vulnerable to theft, blackmail, coersion, and social injustice. It requires us to carefully balance the amount of "smartness" in a system with usable and useful control tools that fit into our social and legal realities.

CHAPTER 5

Supporting Privacy in Mobile and Pervasive Computing

How can future mobile and pervasive computing systems properly take personal privacy concerns and needs into account? How can we ensure that a world full of mobile and pervasive computing will not turn into a dystopian future of mass surveillance? These are hard questions with no simple answers. In this chapter, we discuss key approaches and challenges for respecting and supporting privacy in mobile and pervasive computing. However, none of these approaches alone will "fix" all privacy problems. The previous chapters showed that privacy is a complex topic and protecting privacy in socio-technical systems is a complex challenge. There is no simple technical solution that can fully address all privacy concerns. Nor can technology do so by itself: laws and social norms influence what data practices are deemed acceptable and what should be prevented. Yet, laws and social norms in turn need to be supported by technology, so that they can be implemented in practice. Understanding how mobile and pervasive computing works on a technical level is essential for being able to shape the legal and political privacy discourse, just as it is essential to understand, say, behavioral economics in order to understand decision-making practices of individuals.

The previous chapters provided three key insights.

1. *Privacy provides both individual and societal benefits.* Privacy is not a "nice to have" that we might want to trade in for better shopping experiences or higher efficiency—it is a core requirement of democratic societies that thrive on the individuality of their citizens. Neither is privacy just sought by criminals, delinquents, or deviants, while "honest" citizens have "nothing to hide." Privacy supports human dignity, empowers individuals, and can provide checks and balances on society's powers, such as the government and corporate entities. However, supporting and enforcing privacy comes with costs that individuals often do not or cannot bear—Solove calls this the "consent dilemma" of privacy self-management. We need structural support—legal, technical, and social—in order to enable privacy in mobile and pervasive computing.

2. *Understanding privacy issues and their implications is challenging.* Nissenbaum identifies "contextual integrity violations," Marx describes "privacy border crossings," and Altman describes privacy infringements as a mismatch between an individual's desired and actual "levels of privacy." The "smart" sensing technology at the core of mobile and pervasive

computing is particularly bound to violate those boundaries and levels: it is invisible, large-scale, and inherently data-centric. Mobile and pervasive computing signifies a shift in scale of what information is being collected about individuals, and how long it is being stored. Almost everything is stored permanently by companies and the government. Mobile and pervasive computing often further implies the continuous and implicit collection of information through sensors integrated into personal devices and the environment. Continuous and invisible data collection makes it not only difficult for individuals to know when their activities are being recorded, but also makes it difficult to understand the respective privacy implications, e.g., what higher-level information can be inferred from sensor data. The resulting detailed yet opaque user profiles may lead to discrimination and are prone to manipulation and inaccuracies.

3. *Traditional privacy approaches are difficult to implement in mobile and pervasive computing systems.* Take for example the concept of "notice and choice:" the scale and implicit nature of mobile and pervasive computing systems would place a huge burden on individuals when asking them to individually take a decision on each and every data exchange. However, while new privacy approaches exist, such as "privacy by design," it is not yet clear how these can actually be implemented in mobile and pervasive computing systems.

Following-up on the last point above, the rest of this chapter summarizes seven key approaches for supporting privacy in mobile and pervasive computing: privacy-friendly defaults, integrated privacy risk communication, assisting with privacy management, context-adaptive privacy mechanisms, user-centric privacy controls, algorithmic accountability, and privacy engineering. All of them are active areas of research. We discuss some established solutions, promising directions, and open issues in those areas.

5.1   PRIVACY-FRIENDLY BY DEFAULT

A popular approach to privacy is what Solove [2013] calls *privacy self-management*: through "rights of notice, access, and consent regarding the collection, use, and disclosure of personal data [...] people can decide for themselves how to weigh the costs and benefits of the collection, use, or disclosure of their information." This approach has clearly reached its limits years ago already, even without the additional challenges posed by mobile and pervasive computing. Numerous studies and surveys have shown that there are "severe cognitive problems that undermine privacy self-management" [Solove, 2013]. A study by McDonald and Cranor [2008] estimates that users would need to spend 80–300 hs per year if they were to simply skim each privacy policy they encounter while browsing the Web. Solove [2013] cites work from 2004 that found only 4.5% of users stating they would always read a website's privacy policy [Milne and Culnan, 2004]. These numbers clearly represent a lower bound on the efforts that would be required today, given the proliferation of online services, smartphones and mobile apps over the last decade [Schaub et al., 2017].

Not surprisingly, *default settings* have a significant impact on consumers' privacy choices and their level of protection. According to Acquisti et al. [2017], individuals are subject to a *status quo bias*—most individuals stick with default settings and very few make changes to their privacy settings. Citing Dhingra et al. [2012], they note that people not only have a "propensity to keep the default," but that defaults also influence "which alternatives they choose, should they deviate from the default." Willis [2014] additionally identifies *transaction barriers*, *choice bracketing*, *endorsement effects*, and *endowment effects* as contributing factors in why people stick to defaults. Here, "transaction barriers" refers to the cost of understanding (and executing) the opt-out procedure, potentially repeatedly on a multitude of devices. "Choice bracketing" [Willis, 2014] refers to the fact that an individual decision to share data would be considered trivial to one's privacy, yet the cumulative effect of such sharing decisions could substantially expose private details. The "endorsement effect" describes "the interpretation of a default as a form of implicit advice by a more knowledgeable party as to what most people prefer or ought to prefer" [Willis, 2014]. In behavioral economics, the "endowment effect" further describes the phenomenon that people attribute a higher value to things they already own, i.e., they would sell them for a much higher amount than what they are willing to pay for them if they needed to acquire these in the first place. In a privacy setting, this means that people ask for up to five times as much compensation for giving up presumably private data than they would be willing to pay in order to have otherwise shared data become private.

Whether or not a certain data collection setting is "opt-in" or "opt-out" thus has profound implications for a user's privacy choices. Janger and Schwartz [2002] cite a 2001 survey that found only 0.5% of banking customers had exercised opt-out rights given to them by newly introduced legislation that forced banks to provide such an opt-out choice to their customers.

These findings have two important implications. First, a simple notice-based approach is simply not feasible for users—even more so in a world of mobile and pervasive computing. Second, given the strong bias to keep default settings, in particular for users who are unsure of their own preferences [Acquisti et al., 2015, Willis, 2014], this default choice plays a key role in setting the overall "baseline" of privacy protection.

So what is the correct privacy baseline? As discussed in Section 2.1.2, Europe's General Data Protection Regulation [European Parliament and Council, 2016] explicitly requires "Data Protection by Design and by Default" (Article 25, GDPR):

> The controller shall implement appropriate technical and organizational measures for ensuring that, by default, only personal data which are necessary for each specific purpose of the processing are processed.

Here, "privacy by default" is a specific implementation of the *data minimization* principle, which requires personal data to be "collected for specified, explicit, and legitimate purposes and not processed in a manner that is incompatible with those purposes" and be "adequate, relevant and limited to what is necessary in relation to the purposes for which they are processed" (cf. GDPR Art. 5.1.(b) and (c)).

While the principle of data minimization has received increased attention since the ratification of the GDPR, data minimization is not a new concept. Data minimization was already defined in almost identical wording in Articles 6.1(b) and (c) of the 1995 European Data Protection Directive [European Parliament and Council, 1995] and can be traced back to the "Collection Limitation" and "Use Limitation" principles described in the 1980 OECD guidelines on privacy protection (updated 2013) [OECD, 2013]. Despite this long history, current privacy and data protection regulation is not prescriptive in specifying what these defaults should be—instead the GDPR requires data controllers to only collect what is "adequate and relevant" for the specific purpose of data processing.

Another inspiration for finding the appropriate defaults may come from analyzing the potential of a particular data collection to violate what Nissenbaum [2004] calls "contextual integrity," i.e., the norms and expectations of privacy of a particular context. Nissenbaum's multi-step process, described briefly in Section 2.3.3, focuses on identifying potential violations of an individual's privacy expectations, and thus can help uncover contextually adequate defaults.

Even if personal information is collected adequately, i.e., within the limits of an individual's privacy expectations and "adequate and relevant" for the envisioned purpose of data processing, it is also important to ensure that this data is not used for purposes beyond what it was initially collected for. This so-called concept of "purpose binding" is one of the core principles of the Fair Information Principles (see Section 2.1.1), formulated as part of the 1973 United States Department for Health Education and Welfare (HEW) report [HEW Advisory Committee, 1973]: *"There must be a way for an individual to prevent information about him that was obtained for one purpose from being used or made available for other purposes without his consent."* Today, the concept of purpose binding is an integral part of current privacy legislation (e.g., Article 5.1(b) of the GDPR [European Parliament and Council, 2016] quoted above). Privacy policy specification languages such as P3P [Wenning et al., 2006], EPAL [Ashley et al., 2003], XACML [Parducci et al., 2010], and others [Backes and Markus, 2008] offer an automated way of enforcing such purpose bindings within an information system. However, Koops [2011] questions whether the rigidity of a rule-based system will ever be able to handle the fluidity, flexibility, and context dependency inherent in our legal system, which may help explain the lack of adoption of machine-readable privacy specifications, such as P3P [Cranor, 2012].

5.2   PRIVACY RISK COMMUNICATION

In current privacy regulation and practice, a large emphasis is placed on transparency of data practices. Privacy notices, privacy policies, and mobile permission dialogs aim to make transparent what information about an individual is being collected under what circumstances; how the collected information is used; when and how collected information is shared with other parties; how long it is stored; and privacy controls available to the individual. This notion of transparency is historically rooted in the OECD privacy guidelines [OECD, 2013], the Fair

Information Practice Principles [Federal Trade Commission, 1998, HEW Advisory Committee, 1973], as well as the notion of informational self-determination [Westin, 1967].

Unfortunately, privacy notices often fall short of providing transparency due to long and complex privacy statements [Cate, 2010], which people tend to ignore [McDonald and Cranor, 2008]. Furthermore, the current approach toward privacy transparency often falls short in another crucial aspect: even if individuals are aware of data practices, i.e., what information is collected and why, many struggle to accurately estimate the implications and risks for their personal privacy of those data practices. As we discussed in Chapter 4, seemingly innocuous sensor data, such as heart rate or activity level, can reveal a lot about an individual's habits, health and personality, especially if collected and analyzed continuously. Uncertainty about consequences, framing of privacy-related information, cognitive biases, and decision heuristics detrimentally affect people's privacy decision making [Acquisti and Grossklags, 2005, Acquisti et al., 2015, 2017, Smith et al., 2011]. Behavioral economists refer to this phenomenon of individuals making suboptimal decisions based on incomplete or asymmetric information as *bounded rationality* [Acquisti et al., 2015, Simon, 1982].

Addressing this issue requires us to rethink what constitutes relevant and useful information for people to make effective privacy decisions. Rather than just listing *what* a system's data practices are, privacy notices should emphasize *how* those data practices may affect an individual's privacy. Making privacy risks and implications more explicit would make it easier for indviduals to consider the privacy cost and impact of agreeing to a certain data practice, or when they might want to opt-out or deny a certain practice.

The communication of privacy risks and implications needs to strike a balance between those risks and potential benefits. The goal should not be privacy fearmongering. Instead, a nuanced and balanced presentation of how a specific privacy decision—such as allowing an app access to your location or using a fitness wearable—might impact you, both in terms of direct or intended effects (e.g., the app can provide location-based recommendations; the wearable can track your steps), as well as indirect, less likely or long-term implications (e.g., the app's advertising partners may analyze your location to infer where you live, work and whether you should be targeted with high-priced advertisements or fast food discounts; your fitness wearable could estimate your risk for certain medical conditions and share this information with insurance companies or your employer).

On a more basic level, transparency about privacy risks and implications might help some individuals recognize in the first place that a specific information sharing decision has privacy implications (e.g., that an app shares your location with third-party advertisers). Thus, risk-centric privacy communication can help enhance privacy awareness and support both people with low digital literacy and higher tech saviness.

One way to create awareness for privacy risks is to transparently communicate the implications of a privacy decision at the time individuals are asked to make a decision. For example by laying out what happens if one says "yes" to a sharing decision and what happens if one says "no."

An important aspect of this is to also communicate what happens when one denies a certain data practice. If the app is denied access to your location, does that prevent the app from working? Is the utility degraded? If yes, in what way? Is there a manual option to provide information that would otherwise be collected manually (e.g., selecting a location on a map instead of using location information)? Communicating consequences and options in case of deny or opt-out decisions could help mitigate certain default effects, i.e., the system default being interpreted as a recommendation on what action is preferable [Acquisti et al., 2017, McKenzie et al., 2006].

Furthermore, creating awareness of data practices and potentially associated risks does not only benefit individuals but also the company. Being transparent about data practices and risks helps reduce surprise for users and can help shape a user's understanding and mental model of less expected or seemingly privacy-intrusive practices by explaining the purpose and need for certain data collection [Schaub et al., 2015, 2017]. Thus, effective privacy and risk communication may foster consumer trust in a new mobile and pervasive computing technology.

One area where we are starting to see a privacy communication approach more focused on the consequences of a decision are mobile app permissions. Increasingly, mobile apps sandwich just-in-time permission dialogs (e.g., "This app wants to access your location—Allow / Deny") with additional privacy communication. Figure 5.1 shows an example from Google's Material Design Guidelines for Android permissions [Google, 2017]. Before the system's permission dialog is shown, the app explains why access to the respective permission is needed and sometimes how a "deny" decision might impact app functionality. If the user denies the permission, a subsequent dialog may explain how functionality is now limited, how the user can work around those limitations, or how the user can "allow" access if they change their mind.

Of course, privacy risk communication—if provided by the data processor—might often try to persuade users to choose "allow" and accept a data practice. Yet, the privacy communication around mobile permissions demonstrates that explanations of a privacy decision's consequences can be easily integrated into an application's interaction flow and user experience. What is needed is research, guidance and best practices on how to leverage such opportunities to make privacy risks and implications more transparent. At the same time, there's a need to objectively describe and quantify privacy risks, as well as an opportunity to augment privacy (risk) communication by the data processor with privacy risk communication provided by the system (e.g., the underlying mobile operating system or pervasive computing middleware) or third parties (e.g., researchers, journalists or activists).

5.3   PRIVACY MANAGEMENT ASSISTANCE

The case of mobile app permissions demonstrates another important aspect of privacy: individuals should be given real choices to control their privacy, i.e., they should be able to say "no" to certain data practices, and "yes" to others. For example, a person should be able to agree to getting local weather updates based on their location, while disallowing the detailed analysis of their location data to infer where they live, work, or travel.

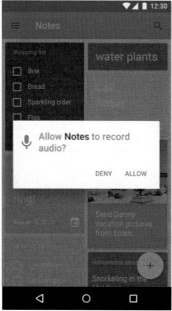
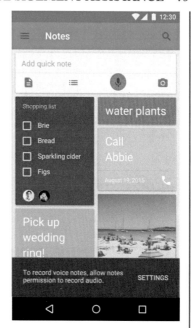

Figure 5.1: An example of privacy explanations around a mobile permission dialog from Google's Material Design Guidelines: an on-screen dialog educates the user why the Notes app wants to record audio (*left*) before the audio recording permission request is shown (*center*); if the user has denied the permission, the app provides information on how to enable the feature if it is desired later on (*right*). Image source: Google [2017].

Many of today's privacy notices do not provide this opportunity. For instance, privacy policies are lengthy documents written in legal language with very little or no choices regarding specific data practices. Individuals are confronted with a take-it-or-leave-it choice: accept the whole privacy policy or do not use the service. In reality, this does not provide a meaningful choice to users [Cate, 2010]. People are forced to accept the privacy policy or terms of service in order to gain access to the desired service—regardless of their actual privacy preferences. If there are no granular choices associated with a privacy notice, individuals have little incentive to read it [Schaub et al., 2017]: it is not worth investing the time to read and understand a privacy policy if there are no real actions one can take, if the policy can change anytime (most privacy policies include a provision to that extent), and if the policy is intentionally abstract and ambiguous about what data practices a user is actually subject to [Bhatia et al., 2016, Reidenberg et al., 2015, 2016]. Abstract and ambiguous descriptions in privacy policies are a consequence of the trend that many companies try to consolidate all their services under a single privacy policy. While this makes it easier for the company to provide its privacy-related information in one place and apply consistent practices across its services, it also means that the privacy policy has to be necessarily abstract and generic in order to cover all services' data collection, use, and sharing practices. As a result, it is often not clear how a specific service covered by the privacy policy actually collects, processes or shares personal information.

Vague privacy notices and policies leave individuals helpless and resigned [Madden, 2014, Turow et al., 2015]. While individuals might care about their privacy, the choice architectures they are presented with force them to accept practices they do not necessarily agree with or are even aware of, because the only other option is to completely abstain from the benefits the service or technology might provide to them [Acquisti et al., 2017, Cate, 2010, Schaub et al., 2017].

The importance of providing actionable privacy information in order to obtain informed consent has been recognized by researchers [Cate, 2010, Cranor, 2012, Schaub et al., 2017] and policy makers [Federal Trade Commission, 2012, 2015, President's Concil of Advisors on Science and Technology, 2014]. Europe's General Data Protection Regulation therefore places strong requirements on how consent has to be obtained, mandating that consent must be freely given, explicit, and specific [European Parliament and Council, 2016]. However, the challenge is to provide individuals with real and granular privacy choices without overwhelming them with those choices. Just exposing more and more opt-ins, opt-outs, and privacy settings to users will not scale, as it will just overwhelm individuals with choices and place the burden for privacy management solely on them [Solove, 2013]. Obtaining meaningful consent without overwhelming users is particularly challenging in mobile and pervasive computing, given the often implicit and imperceptible nature of data collection via sensors embedded into devices and the environment [Luger and Rodden, 2013b], as well as the fact that privacy is commonly not the user's primary task or motivation in using a system [Ackermann and Mainwaring, 2005].

For instance, mobile permissions already require individuals to make dozens if not hundreds of privacy decisions for all the resources the apps on their smartphones want to access. Based on a sample of 3,760 Android users, Malmi and Weber [2016] find that Android users in 2015 had used 83 smartphone apps at least once per month, on average. Given that on average each app requests five permissions [Pew Research Center, 2015], this results in over 400 permission decisions a smartphone user would have to make on average. Expanding the smartphone permission approach to online services, wearables, smart home systems and other pervasive computing systems, as well as all data collection, use, sharing, and retention practices would result in a plethora of privacy settings, choices, and permissions. Even the most interested user would not be able to consistently manage their privacy across all those choices and contexts [Liu et al., 2016, Schaub, 2014, Schaub et al., 2015], as well as over time [Luger and Rodden, 2013a].

A potential way forward are privacy management assistance solutions that aim to help users more effectively manage their privacy. For instance, machine learning can be leveraged to create personalized privacy assistants or privacy agents which learn from an individual's privacy decisions and either provide recommendations for future decisions in the same context or across contexts or even automate privacy decisions for the user [Kelley et al., 2008, Knijnenburg and

Kobsa, 2013, Liu et al., 2016, Sadeh et al., 2009, Schaub et al., 2015]. The personalized pre-diction of privacy settings and preferences has received considerable attention [Cornwell et al., 2007, Cranshaw et al., 2011, Kelley et al., 2008, Lin et al., 2012, 2014, Sadeh et al., 2009]. There are two general approaches: leveraging an individual's prior privacy decisions to predict preferences in new decision contexts or assigning an individual to a *profile*, a cluster of people with similar privacy preferences. Such profiles could be based on clustering privacy settings of other users, learning from privacy settings made by the user's "friends," or be curated by ex-perts [Agarwal and Hall, 2013, Toch, 2014]. These two approaches can also be combined: an individual can be assigned to a profile initially to bootstrap a privacy assistant and avoid coldstart issues, followed by experiential refinement and privacy preference learning from an individual's actions over time [Knijnenburg and Kobsa, 2013, Schaub, 2014, Schaub et al., 2015]. For exam-ple, through a set of questions, the system might learn that a person is highly concerned about location data. By assigning the person to the "location privacy" cluster, the privacy assistant may determine that for this person location sharing should either be blocked by default or the user should be asked when location is being accessed. Based on the person's privacy settings adjust-ments over time, the assistant might learn that while the user prefers not to share location in general, the user regularly grants location access to transportation-related apps (e.g., navigation, ride sharing, bus schedule) and update the user's privacy preference profile accordingly.

Reasoning results can either be provided as privacy settings recommendations to users or settings can be automatically adjusted for the user. Typically, the level of confidence in the reasoning result determines the level of automation or user involvement in making the privacy decisions [Bilogrevic et al., 2013, Kelley et al., 2008, Knijnenburg and Kobsa, 2013, Schaub, 2014, Schaub et al., 2015, Toch, 2011]. Determining the appropriate level of automation—with multiple stages between fully manual and fully automated configuration [Parasuraman et al., 2000]—for privacy decision making and privacy management assistance is a topic of active re-search. Enabling auditing of system decisions and giving users controls to adjust and tweak preference models, for instance through overviews of what apps were allowed or denied access to location and other resources [Schaub et al., 2014, Tsai et al., 2017], both improves the privacy assistant and provides the users with agency over their privacy assistant.

While researchers have made progress in accurately predicting individuals' preferred pri-vacy settings in specific situations, it often remains a challenge to help individuals make those settings or automate settings changes for them. Current privacy assistance solutions are either confined to their application context, in which they learn from users' behavior within the ap-plication and may adjust recommendations and settings within that context [Das et al., 2018, Knijnenburg and Kobsa, 2013, Schaub et al., 2014], or require modifications to the underlying system layer, e.g., research prototypes with superuser rights hooking into Android's permission framework [Liu et al., 2016, Wijesekera et al., 2018].

There is a call for the need to expose privacy settings APIs which would allow privacy as-sistants to function across contexts [Liu et al., 2016], positing that otherwise we will end up with siloed and less useful privacy assistants for different systems and services, such as Facebook, your smartphone apps, or (parts of) your smart home. Such disjoint privacy assistants would not be able to leverage and benefit from privacy decisions made in other contexts to improve prediction accuracy, requiring the re-learning of privacy preferences in different application contexts and for different privacy assistants. However, given the context-dependent nature of privacy perceptions and preferences [Acquisti et al., 2015, Nissenbaum, 2009], highly specialized privacy assistants might be able to provide more relevant support in the context they are focusing on (e.g., smartphone apps) than general privacy assistants that aim to comprehensively model an individual's privacy preferences, which may be fraught with uncertainty and context dependency as behavioral economics research suggests [Acquisti et al., 2015]. Nevertheless, further research on more deeply understanding factors that affect privacy decisions and modeling privacy decision making processes is essential to further improve privacy management support.

5.4 CONTEXT-ADAPTIVE PRIVACY MECHANISMS

A particular challenge in developing "smart" privacy assistants and controls for mobile and pervasive technologies is the dynamic nature of both privacy preferences and privacy implications [Acquisti et al., 2015, Nissenbaum, 2009, 2011]. Mobile and pervasive computing exacerbate those aspects due to the inherent and often continuous collection of sensor and context data enabling further inferences about an individual's behavior, as we discussed in Chapter 4. Privacy solutions need to account for the context-dependent nature of privacy needs, otherwise they may shine in a constrained context but may falter in another. Adapting privacy assistance solutions to a different context or application domain often requires complete re-training and reconsideration to fit the approach to the new circumstances.

Context can affect and shape privacy expectations and needs on multiple levels [Nissenbaum, 2009]. Privacy expectations, preferences, and needs may differ across *application domains*. For example, physiological data streams may be acceptable and desired for fitness and health—both for personal use and in collective settings—but may create privacy concerns in employment and health insurance contexts, where activity data could be used to estimate work performance, insurability and risk factors. Expectations and the conceptualization of privacy are also shaped by *sociocultural context*. What may be considered an invasion of privacy in one culture may be acceptable in another; or what may be considered common privacy-seeking behavior in one cultural context, may be less acceptable in another. For example, in the Netherlands most homes do not have window curtains or blinds and curtains are rarely closed, whereas it is common in other cultures to close curtains to create a more private sphere for domestic life.

For individuals, privacy preferences, expectations, concerns, and requirements may also change over *time*. Personal experiences, knowledge gained, or changes in circumstances affect the awareness of privacy risks and the needs for privacy. Privacy perceptions and expectations may also change on a societal level over time. For instance, when new privacy-related knowledge pervades the collective consciousness. A recent example are Edward Snowden's revelations about large-scale and persistent monitoring and analysis of personal internet communication and other digital traces by intelligence agencies from the U.S. and other countries [Landau, 2013, 2014, Madden, 2014]. Whereas comprehensive surveillance of online behavior was considered a possibility by security experts, it is now established fact that intelligence agencies gather and analyze digital traces of large parts of the population. This knowledge can result in chilling effects on the expression of government-critical sentiments or engagement in controversial activities. Indeed, studies have shown chilling effects on online searches after the Snowden revelations: searches for privacy-sensitive terms declined significantly both in Wikipedia [Penney, 2016] and Google Search [Marthews and Tucker, 2017] after the extent of U.S. government surveillance became public. Another example is the collective realization, spurred by the debate about Russian meddling in the 2016 U.S. election, that behavioral information gathered by Facebook and other online services can be used to predict personality traits and can be exploited to influence political sentiments and even elections. Privacy experts had warned about how revealing online behavior can be for many years, but the 2016 U.S. election as well as the news that a third party, Cambridge Analytica, had obtained large amounts of Facebook user data [Valdez, 2018] transferred this knowledge into the public consciousness.

Contextual factors influencing privacy can be manifold. Considering and anticipating privacy-relevant context aspects *a priori* in the design of privacy mechanisms can be challenging. Rather than requiring manual adjustment and tuning to each new context, privacy mechanisms can be designed to dynamically adapt to changes in context. Privacy mechanisms and privacy assistants should be able to dynamically adapt to certain context changes by accounting for context in how privacy assistants function. Furthermore, privacy assistants need to provide tools for individuals to reflect on their privacy settings and preferences in certain contexts and support them to make adjustments.

While the inherent context awareness of mobile and pervasive computing systems poses many privacy challenges, this context awareness also provides an opportunity for privacy in enabling such context-adaptive privacy mechanisms [Schaub, 2018, Schaub et al., 2015]. Similar to how contextual aspects play into how an individual makes privacy decisions [Altman, 1975, Nissenbaum, 2009], a context-adaptive privacy system can analyze and react to privacy-relevant changes in context in order to support its users in managing their privacy across changing contexts. Schaub et al. [2015] operationalize this approach by structuring how context-adaptive privacy mechanisms function into three consecutive phases: *privacy context modeling* to provide situational privacy awareness, *privacy reasoning and adaption* to support privacy decision making, and *privacy preference realization* to translate a privacy decision into actual privacy configurations in a given context and environment.

Privacy context modeling requires an understanding of how and at what granularity privacy-relevant context factors can and need to be represented in computational context models. For instance, Palen and Dourish [2003], Lehikoinen et al. [2008], and Romero et al. [2013] applied Altman's boundary regulation theory to pervasive computing in order to identify situations in which information exposure and privacy implications change. Furthermore, privacy-relevant context information needs to be collected either by deriving privacy-relevant context features from sensor information, e.g., by sensing people nearby [Könings et al., 2016, Schaub et al., 2014], or by communicating privacy information in machine-readable formats. For instance, Langheinrich [2002a] early on proposed a privacy awareness system (pawS), in which *privacy beacons* send out privacy policy information encoded with the P3P standard [Wenning et al., 2006] which is received by the user's *privacy proxy* and matches a system's data practices against the user's privacy preferences. This approach has been adopted and expanded to facilitate context-based privacy negotiation in pervasive and IoT environments [Das et al., 2018, Kwon, 2010, Yee, 2006], as well as enable users to communicate their preferences to a pervasive environment, e.g., through a mobile app sending out privacy preference beacons [Könings et al., 2014].

When privacy-relevant context changes are detected, context-adaptive privacy mechanisms need to match or predict what the user's privacy preferences may be for the new situation. Reasoning about the user's privacy preferences determines whether a user should be warned or informed about the change, whether privacy settings can be adapted to the new situation automatically, or whether a set of recommended actions should be provided to the user. Privacy systems might be rule-based (e.g., "clear this display when a new person is detected"), for instance assigning disclosure rules to physical objects [Roesner et al., 2014] or adjusting displayed content based on context rules [Davies et al., 2014, Könings et al., 2016, Schaub et al., 2014]. While effective for specific contexts, rule-based approaches struggle with the uncertainty and fuzziness in both context detection and user's privacy preferences in substantial context changes. More advanced approaches rely on case-based reasoning leveraging prior decisions of an individual [Bernsmed et al., 2012, Sadeh et al., 2009, Saleh et al., 2007, Schaub et al., 2014] or extract privacy preference profiles from a large number of individuals' privacy settings [Knijnenburg and Kobsa, 2013, Liu et al., 2016] or their expressions of privacy preferences and concerns [Lin et al., 2014, Wijesekera et al., 2017].

All these approaches require substantial amounts of privacy preference or behavioral data. In collecting such data, especially when eliciting privacy preferences and concerns through self reports, one has to be cautious of the privacy paradox [Norberg et al., 2007]: people's stated privacy preferences and concerns are not always reflected in their actual behavior. Stated privacy preferences and concerns are usually aspirational—they describe a desired level of privacy, whereas actual behavior is affected by uncertainty, cognitive biases, decision heuristics, and is malleable by framing and offered choice architectures [Acquisti et al., 2015, 2017]. Context-adaptive privacy reasoning approaches should account for uncertainty about sensed context as well as a user's privacy preferences. Similar to privacy assistance solutions in general, the level of confidence in the context-specific privacy reasoning outcome should determine the required level of human participation in an adaptation decision: systems might automatically adapt privacy settings in high confidence cases with or without informing the user, whereas the user might be offered a set of recommended privacy actions in lower confidence cases [Schaub et al., 2015]. The ten levels of human interaction with automation by Parasuraman et al. [2000] can help guide the appropriate level of user involvement in the privacy adaption.

A particular challenge is enabling context-adaptive privacy mechanisms to effectively realize adaptation decisions in heterogeneous environments. Different systems, infrastructure, sensors, services, or devices may be controlled by different stakeholders [Ackerman and Darrell, 2001]. How a privacy reasoning result can be realized in a specific environment depends on the privacy control capabilities available, i.e., the level of control and trust concerning other devices, infrastructure, and entities [Könings and Schaub, 2011]. Context-adaptive privacy mechanisms may have to consider different adaptation strategies within the same context as well as across contexts and systems. Furthermore, privacy controls available within a specific environment should be discoverable in order to facilitate privacy management and adaptation in previously unknown environments and contexts [Das et al., 2018, Langheinrich, 2002a].

5.5   USER-CENTRIC PRIVACY CONTROLS

On a more fundamental level, we also need to rethink *how* privacy settings and controls are structured on a systems level and how they are made available to users. Current privacy controls are rooted in access control. Almost always privacy settings are binary access decisions. Once access to a resource has been granted to an app or system, the resource can be used without further restriction until access is revoked again. For example, once you grant a mobile app access to your location, the app can access location information whenever it wants to and for whatever purpose until the permission is revoked. This is inconsistent with both how people make privacy decisions and certain legal requirements (e.g., GDPR [European Parliament and Council, 2016]) and privacy protection principles (e.g., the OECD's purpose specification and use limitation principles [OECD, 2013]) prescribing that consent for data processing has to be limited to a specified purpose. Context information can be used to infer and contrain the legitimacy of mobile apps' data access [Chitkara et al., 2017, Tsai et al., 2017, Wijesekera et al., 2017]. New data and permission management approaches, such as PrivacyStreams [Li et al., 2017], aim to change the underlying permission model to make it easier to constrain and audit data flows.

Furthermore, continuous data collection through sensor streams creates a need for potentially excluding certain moments or situations from collection. For instance, user should be able to *proactively* pause collection, e.g., for a certain period of time, adjust the *granularity* of collection, or *retroactively* remove collected data. Such controls could both be available for manual control by the user or through automated support. For example, as discussed in Section 5.4, systems could detect potentially sensitive contexts and suggest to the user when to "tighten" privacy settings, as well as when to "loosen" them again. There is substantial opportunity for innovation in designing privacy controls that facilitate more nuanced decisions but are also easy to use. Such controls also provide an interesting opportunity for integrating privacy risk communication by estimating privacy implications of data collection in specific contexts and automatically flagging sensitive or interesting situations the user might have different privacy preferences for.

In mobile and pervasive systems, users should also be given opportunities to reflect on and adjust privacy settings over time. Privacy settings and retroactive privacy controls in current systems typically give users the option to revoke access permissions and limit future collection; data access tools enable users to review and possibly remove collected data. What requires more attention is how to design such retroactive privacy controls in a way that supports users in reverting previous privacy decisions, i.e., if a user changes their mind about a permission previously granted, they may be able to relatively easily deny or limit future data processing, but it is often difficult and tedious to revoke permissions or consent for data previously collected. What is missing are usable rollback solutions that support users in retroactively and selectively limiting the use of previously collected data. Such approaches should not just support deletion of directly collected data but extend to any inferences made from that data, as well as facilitate distributed data deletion and revocation of consent for data that has been shared with other entities. Such retroactive privacy controls are gaining in importance due to the GDPR's right to erasure, also called the "right to be forgotten" (Article 17, GDPR). Considering and investing in the usability of such mechanisms will also benefit system developers because it may allow for more differentiated and selective data deletion by users than a simplistic "delete my account and all data" option.

Another worthwhile consideration is whether users can be not only supported to delete their data but also retroactively share more data. If at some point the user decides to activate or reactivate data collection that was previously denied, can—if so desired—past information be shared? While this may sound counterintuitive—how can you share information that hasn't been collected?—middleware solutions for context-aware computing increasingly act as gate keepers for context or sensor information. For example, Apple's iOS devices continuously collect activity data from motion sensors and, through the Apple Health API, apps can request access to both future and past sensor information. However, those solutions are currently still binary—either you grant access to *all* past and future data or to nothing. Yet, context awareness provides opportunities to group and categorize sensor data into events, time spans and situations which may allow for more specific and more intuitive sharing settings.

Overall, context awareness does not just pose privacy challenges but also provides exciting opportunities to design and develop novel privacy controls that are better aligned with people's contextualized privacy preferences, while also facilitating better integration of privacy controls into the user experience of a mobile or pervasive system. Current privacy controls are often hidden in privacy settings disconnected from the primary user experience, and are therefore under-utilized. Context-aware privacy mechanisms can be better integrated into the user's interaction flow with the system without disrupting the user experience by providing information and controls relevant to the user's current context and activity.

5.6 ALGORITHMIC ACCOUNTABILITY

One of the key issues of building comprehensive data collections on individuals is the risk of *profiling* [Hildebrandt, 2008, Schermer, 2011], as we discussed in Section 4.3. Profiling can be defined as an act of *algorithmic decision making* on user profiles. Based on collected personal data, an algorithm is responsible for the *prioritization* and ultimately *filtering* of people (e.g., applicants), the *association* of people into similar groups, and ultimately their *classification* into categories (e.g., as preferred customer) [Diakopoulos, 2016].

At the outset, an individual profile may be inaccurate and thus prevent its subject (i.e., the person the profile is about) from receiving certain benefits they would otherwise receive. Unfortunately, there are many real-world examples for the dangers of relying on algorithmic decision making. In 2013, the Michigan Unemployment Insurance Agency (UIA) replaced a 30-year old mainframe system for deciding unemployment benefits. The new system, dubbed MiDAS (Michigan Integrated Data Automated System), quickly seemed to pay off: in its first few weeks of operation, MiDAS not only uncovered a fivefold increase in fraudulent unemployment filings, but also triggered fraud proceedings against the offenders without any need for human intervention (eliminating one third of UIA's staff in the process). Less than two years later, and only after coming under immense public pressure, UIA halted this automated fraud assessment program: it turned out that over 92% of its automated decisions were wrong [Charette, 2018]. Nearly 21,000 individuals not only had significant parts of their income and/or assets wrongfully seized by the state, but were also criminalized by a rogue computer program without any information on what precisely they had done wrong, nor were given the chance to defend themselves. Charette [2018] provides a list of other such cases (e.g., Australia's Centerlink benefits system).

Even an accurate profile can be problematic, as it may expose its subject to the threat of identity theft due to it being a comprehensive collection of private information about a person, which an attacker could use to impersonate that individual. Profiles that are applied to groups (e.g., to people from a certain geographic region) may equally entail individual disadvantages, as group attributes are uniformly applied to all group members (e.g., expectation of higher crime rate). Thus, profiles ultimately carry also societal risk, as their widespread use makes identity theft more likely and can contribute to discrimination in society.

At the same time, some form of "profiling" and algorithmic decision making are integral for the functionality of many mobile and pervasive computing services, ranging from recommending locations and stores near-by that are relevant to the user to automatically adjusting a home's temperature to the preferences of those present. Therefore, privacy-aware systems will need to find a way to minimize the negative impacts of profiling and algorithmic decision making. This can be achieved using three main approaches: transparency, redress, and data minimization.

At the outset, individuals subject to an automated decision taken on the basis of an individual profile should be made aware of this process. Without this baseline level of transparency, an individual might not even be aware of any disadvantages their profile might cause them. The dangers of automated profiling already feature prominently in European data protection law. The General Data Privacy Regulation [European Parliament and Council, 2016] states in Article 22 on "automated individual decision making, including profiling" that individuals *have the right not to be subject to a decision based solely on automated processing, including profiling, which produces legal effects concerning him or her or similarly significantly affects him or her*" except when necessary for a contract, required by law, or with explicit consent. Further restrictions apply if sensitive data (as defined in Article 9, GDPR) are being processed. Furthermore, the GDPR requires data controllers to "*provide the data subject with … information*" about "*the existence of automated decision-making, including profiling, referred to in Article 22(1) and (4) and, at least in those cases, meaningful information about the logic involved, as well as the significance and the envisaged consequences of such processing for the data subject*" (GDPR Art. 13.2(f)).

Providing meaningful transparency is not trivial, given the complexity and sophistication of today's profiling and machine intelligence algorithms [Knight, 2017]. At the outset, this requires informing subjects about the classification results of the system (e.g., "high-income spender," "credit risk"), though such human-readable descriptions seem increasingly quaint when more and more processing is directly handled by self-learning systems: a classification system may dynamically generate such classes of users and label them in a generic fashion that may be difficult (if not impossible) to map onto meaningful human descriptions. Secondly, the actual reasons for classification should be communicated, yet again the sheer number of data points used for today's profiling systems makes this difficult to achieve in practice. For example, research has been able to infer sexual orientation from one's social network [Jernigan and Mistree, 2009], Facebook "Likes" [Chen et al., 2016], or profile pictures [Wang and Kosinski, 2018]—hence communicating to a data subject why, for example, their profile picture causes them to fall in one category or another would at best mean disclosing the trained weights that led the employed neural network to come up with its decision. This would hardly be "transparent" for most people.

The second approach, redress, clearly relies on sufficiently transparent processes in order to offer meaningful ways for data subjects to seek help in case of misclassified information. If one does not even know that a certain outcome (e.g., a denied loan) is the result of an automated assessment based on a (potentially) inaccurate profile, seeking redress is difficult. Crawford and Schultz [2014] propose to provide citizens with a new legal right—*procedural data due process*. Similar to existing *procedural due process* rights, which give individuals certain rights to review and contest the evidence against them, the authors propose that companies would need to allow individuals to "review and contest" the profile and data upon which an automated decision was based. McClurg [2003] argues for a new approach to U.S. tort law (corresponding to the civil law legal system in contiental Europe) that would make the unauthorized selling of consumer profiles by data brokers actionable under an *appropriation* tort.[1]

Others have proposed technical approaches that would allow data processors engaging in data mining to minimize the risk of their algorithms violating anti-discrimination or privacy laws, e.g., by flagging sensitive attributes [Fule and Roddick, 2004], replacing discriminative factors (e.g., ethnicity) with implicitly present factors (e.g., education) [Schermer, 2011], or by detecting rules *ex post* that produce significantly different outcomes for particular subgroups of a population [Ruggieri et al., 2010]. Note that simply adding human oversight to intransparent systems alone is not a sufficient solution. As Cuéllar [2016] points out, even systems that incorporate human review are seldom challenged in their automatically generated decisions, given the often-assumed "objectivity" of computer algorithms [Marx, 2003].

Data minimization, the third approach for addressing potential negative aspects of profiling, has been a tenet of privacy laws. Already the 1980 OECD guidelines [OECD, 1980] stipulated Data Quality as its second principle: *"Personal data should be relevant to the purposes for which they are to be used"*. The 1995 EU Privacy Directive [European Parliament and Council, 1995] was much more explicit, asking in its Article 6.1(c) that the collected that was not only relevant, but also *"not excessive in relation to the purposes for which they are collected and/or further processed"*. The same principle, sometimes called the "proportionality and necessity principle," can now be found in Article 5.1(c) of the GDPR [European Parliament and Council, 2016] in slightly revised form: personal data shall be *"adequate, relevant and limited to what is necessary in relation to the purposes for which they are processed ('data minimization')."* As we pointed out before, determining what data collection and processing is "relevant" and "not excessive" is far from trivial—in particular when collected with the intention of building a general profile of a data subject. It is worth noting that the principle applies to both the collection *and* the processing of personal data, highlighting the fact that data, once collected, may see multiple "uses" when a profile is used to infer a range of attributes.

Several authors have pointed out that "classic" privacy principles such as data minimization and consent may not be adequate for systems that rely heavily on profiles and data mining [Rubinstein, 2013, Tene and Polonetsky, 2012, 2013]. Instead, systems should favor accountability principles such as transparency, access, and accuracy:

> The shift is from empowering individuals at the point of information collection, which traditionally revolved around opting into or out of seldom read, much less understood corporate privacy policies, to allowing them to engage with and benefit from information already collected, thereby harnessing big data for their own personal usage [Tene and Polonetsky, 2013].

---

[1]An appropriation tort provides for "liability against one who appropriates the identity of another for his own benefit" [McClurg, 2003]. It is typically used by celebrities for safeguarding their "brand value" when others try to sell a product or service based on the celebrity's likeness (e.g., image).

It is obvious that data that has not been collected (data *collection* minimization) cannot be misused. However, given the likelihood that data processors will incentivize data subjects to consent to broad data collection practices (and thus will be able to build comprehensive profiles), ensuring subject-controlled use of the data (data *processing* minimization) is critical.

5.7 PRIVACY ENGINEERING

A pre-requisite to most privacy approaches discussed so far is a system's ability to *support* such privacy-friendly behavior in the first place. Thus, an important question is how to effectively integrate privacy-friendly features into a system.

The systematic approach to such an integration is called *privacy engineering*. Privacy engineering comprises the "design, implementation and integration of privacy protection" in products and services [Gürses et al., 2015]. This goes beyond what is typically called "privacy-enhancing technologies" (PETs)—specialized privacy techniques or mechanisms that address a specific privacy problem. Those PETs—while often powerful—do not make it into the majority of products due to associated technical complexity. Privacy engineering thus is not only about technology, but about a *process* to integrate privacy protections into actual systems. Spiekermann and Cranor [2009] define two key aspects of such a system privacy integration:

1. ensuring that users can exercise immediate control over who can access their personal data (or, alternatively, who can access the person themselves, e.g., with a call or a message); and

2. ensuring that personal data, once disclosed, is protected so as to minimize subsequent privacy risks.

Depending on the actual system, these two functionalities can be located across three "domains" [Spiekermann and Cranor, 2009].

1. The *user sphere*: data located on a user's device, or—in the case of physical privacy—the reachability of a user through this device.

2. The *recipient sphere*: data located in a third party's backend infrastructure.

3. The *joint sphere*: data directly accessible to a user but technically hosted on third party infrastructure (often free of charge), e.g., free Web-based email or cloud services.

At the outset, privacy engineering seeks to enumerate protection goals, identify threats and vulnerabilities, and catalog technical solutions that fit the corresponding requirements in order to support engineers in building privacy-aware systems across IT processes. However, to be effective, privacy engineering also must include an understanding of end-user behavior and expectations, in order to provide the right level of support at the right time.

Spiekermann and Cranor [2009] point out that one can either integrate privacy protection into a system's software architecture (*privacy by architecture*), or use legal and organizational means (*privacy by policy*) to ensure such protection. Privacy-by-architecture approaches represent *protection services*, usually cryptographic algorithms, such as zero-knowledge proofs, that prevent adversaries from making unintended inferences, as well as appropriate access control and security. Privacy-by-policy approaches represent *deterrence services* that use threats of disciplinary or legal consequences if individuals and organizations do not follow accepted or agreed-upon practices for the collection and use of personal data. Ideally, such privacy-by-policy approaches stipulate the creation of respective internal privacy policies, practices and processes within organizations.

Gürses and del Alamo [2016] add a third approach, dubbed *privacy by interaction*, which situates privacy supporting mechanisms at the socio-technical level, i.e., directly addressing the conceptualization of privacy as a contextual process [Nissenbaum, 2009]. When seen as a security service, privacy-by-interaction approaches work as *resilience services* that lower the impact of a data breach by virtue of taking user concerns and wide-scale impacts into account when designing such services in order to limit potential exposure of personal information. Gürses and del Alamo [2016] point out that successful privacy engineering requires all three of these approaches to be applied together, rather than individually (e.g., software developers addressing architectural problems, lawyers considering policy issues and UX designers investigating interaction challenges).

How system design and development can best integrate these three approaches is an area of active research and emerging professional practice. The basic idea of this approach—considering and integrating privacy at the outset of a system's design—was suggested by Cavoukian [2009] with the concept of *Privacy by Design* (PbD). While Cavoukian's principles suggest consideration of privacy at a high-level (e.g., "retain full functionality" and "respect user privacy"), they lacked concrete privacy-preserving tools and techniques as well as a systematic methodology to integrate them into a system's lifecycle. The concept of privacy by design, however, has since been integrated into EU privacy law, notably Article 25 of the GDPR [European Parliament and Council, 2016] called "Data Protection by Design and by Default," which requires data controllers to *"implement appropriate technical and organizational measures […] designed to implement data-protection principles […] in an effective manner, and to integrate the necessary safeguards into the processing […]."*

Danezis et al. [2015] offer eight practical privacy "design strategies," derived from Hoepman [2012], as summarized in Table 5.1. The first four strategies—MINIMIZE, HIDE, SEPARATE, and AGGREGATE—are called "data-oriented strategies," roughly corresponding to the concept of *privacy-by-architecture* and implementing data minimization. The latter four strategies—INFORM, CONTROL, ENFORCE, and DEMONSTRATE—are called "process-oriented strategies" and correspond to both the concepts of *privacy-by-policy* and *privacy-by-interaction*.

While these strategies offer a more operationalizable approach to privacy-by-design—in particular when combined with illustrative examples of technical and methodological approaches

Table 5.1: Design strategies for implementing privacy-by-design [Danezis et al., 2015]

#	PbD Strategy	Description
1	MINIMIZE	The amount of personal data that is processed should be restricted to the minimal amount possible.
2	HIDE	Personal data, and their interrelationships, should be hidden from plain view.
3	SEPARATE	Personal data should be processed in a distributed fashion, in separate compartments whenever possible.
4	AGGREGATE	Personal data should be processed at the highest level of aggregation and with the least possible detail in which it is (still) useful.
5	INFORM	Data subjects should be adequately informed whenever personal data is processed.
6	CONTROL	Data subjects should be provided agency over the processing of their personal data.
7	ENFORCE	A privacy policy compatible with legal requirements should be in place and should be enforced.
8	DEMONSTRATE	A data controller should be able to demonstrate compliance with the privacy policy and any applicable legal requirements.

of each strategy, as provided by Danezis et al. [2015]—they still do not fully address how to integrate privacy engineering in the overall system design process. Kroener and Wright [2014] propose to combine best-practices, privacy-enhancing technologies (PETs), and privacy impact assessments (PIAs) in order to operationalize PbD.

According to Wright [2011] PIAs started to emerge since the mid-1990s in countries around the world. In Europe, the UK privacy commissioner, the Information Commissioner's Office (ICO), published the first handbook on privacy impact assessments in 2007 [ICO, 2007]. The handbook saw a first revision in 2009 and has since been integrated into the more general Guide to the General Data Protection Regulation (GDPR) [ICO, 2018], which the ICO regularly updates. PIAs draw on the tradition of environmental impact assessments and technology assessments, which have a long tradition in both the U.S. and Europe. Clarke [2009] offers a comprehensive history and Wright [2011] provides an overview of PIA requirements around the world, as well as a detailed description of typical PIA steps [Wright, 2013].

PIAs set out a formal process for identifying (and minimizing) privacy risks by drawing on input from all stakeholders. The key steps are [ICO, 2018]:

1. identify the need for a PIA;

2. describe the information flows;

3. identify the privacy and related risks;

4. identify and evaluate the privacy solutions;

5. sign off and record the PIA outcomes;

6. integrate the outcomes into the project plan; and

7. consult with internal and external stakeholders as needed throughout the process.

PIAs have since become a mandatory part of privacy legislation in Europe, the U.S., and other countries. For example, the GDPR [European Parliament and Council, 2016] requires a "data protection impact assessment" (DPIA) in Article 35, at least for processing that "is likely to result in a high risk to the rights and freedoms of natural persons." While the GDPR does not mandate a formal methodology for conducting such a DPIA, its expected outcomes largely overlap with those of a PIA. Several authors offer advice on how to integrate PIAs into system development, e.g., Kroener and Wright [2014] and Oetzel and Spiekermann [2014]. In the United States, Section 208 of the E-Government Act of 2002 requires government agencies to conduct and publish Privacy Impact Assessments when developing or procuring information technology for the collection and processing of personally identifiable information, or when considering new data collection processes [United States Government, 2002]. Similar laws with PIA requirements exist in many other countries.

A more comprehensive privacy engineering methodology was proposed by Notario et al. [2015] as the outcome of a research project called "PRIPARE—Preparing Industry to Privacy by Design by supporting its Application in Research." The authors propose a two-pronged approach that uses both risk-based privacy requirements analysis and goal-based privacy requirements analysis (see Figure 5.2). The authors map PIAs onto the risk-based requirements analysis, while legal requirements and codes-of-conduct are part of a goal-oriented requirements gathering process. Both sets of requirements can then drive a design methodology using either a top-down approach that instantiates an architecture from those requirements (e.g., following the CAPRIV framework), a bottom-up approach that works off existing code, or a "horizontal" approach (e.g., CBAM [Nord et al., 2003] or ATAM [Kazman et al., 1998]) that starts with an initial architecture and iteratively refines them based on the identified privacy requirements.

These approaches are just examples of the many methodologies for integrating privacy into the engineering process. However, this multitude does not mean that this is easy, but rather that the best solution has yet to be found. For current research on privacy engineering, the *Workshop on Privacy Engineering (IWPE)* series[2] might offer a good starting point. In 2018, the International Association of Privacy Professionals (IAPP) also established a Privacy Engineering Section, dedicated to advancing privacy engineering practice.[3]

---

[2]Workshop on Privacy Engineering Series: http://iwpe.info/.
[3]IAPP Privacy Engineering Sec.: https://iapp.org/connect/communities/sections/privacy-engineering/.

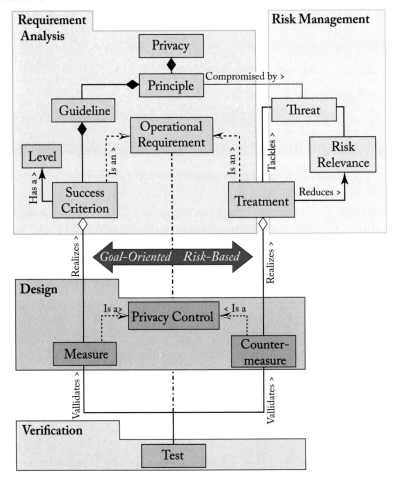

Figure 5.2: The PRIPARE Privacy-by-Design Methodology considers both goal-based and risk-based privacy requirement analyses (base on Notario et al. [2015]).

5.8   SUMMARY

In this chapter, we provided an overview of paradigms, mechanisms and techniques to design and develop privacy-friendly mobile and pervasive computing systems. While there's no silver bullet to remedy *all* privacy implications of *any* mobile and pervasive system, the presented approaches constitute an essential toolbox for building privacy into mobile and pervasive computing systems. An important first step for any system is the integration of privacy-friendly defaults. Data collection and use should be minimal—no more than necessary for the system to function—and potentially privacy-invasive or unexpected data practices should not be active by default but rather be activated by the user in order to help users in constructing a more accurate mental model of what data a system collects and how that data is used. Just-in-time mobile permission dialogs are a common example of this approach.

Rather than just informing users about data practices, associated privacy risks and implications should further be communicated to users at such occasions, in order to minimize surprise and help users make meaningful decisions regarding their privacy in the use of mobile and pervasive technologies. Furthermore, provided privacy information should always be actionable in order to assist individuals both in forming and expressing privacy decisions through privacy settings and options. At the same time, individuals are faced with an increasing number of privacy decisions. Machine learning can help provide personalized privacy assistance by learning from people's privacy preferences and recommending likely preferred settings to individual users. Context-adaptive privacy mechanisms extend this notion to providing context-specific privacy assistance by leveraging context-aware sensing infrastructure and computing approaches to dynamically adjust privacy settings to context changes.

A privacy challenge that arises with automated systems, is that algorithmic decision making can incidentally codify discriminatory practices and behavior in algorithms. Thus, a substantial challenge for artificial intelligence research is providing transparency about how automated decisions are made in an intelligible and correctible fashion. Designers and developers of mobile and pervasive computing systems—which derive much of their benefits from ingesting sensor information and inferring state and required actions—need to consider how they can detect and eliminate undue bias in their decision engines, provide algorithmic transparency to users, as well as offer opportunities for correcting errors as well as provide redress processes for harmed users.

The privacy challenges and approaches discussed put a lot on the plate of a designer or developer of a mobile or pervasive system or application. Engaging in privacy engineering and considering privacy from the beginning, can help surface privacy issues and challenges early-on in the design, as well as facilitate the consideration of creative privacy approaches and solutions. Similar to other engineering practices and considerations, relying on established frameworks, such as privacy impact assessments, can provide structured guidance in the process and enable reliable and consistent privacy outcomes.

CHAPTER 6

Conclusions

Many privacy experts cheered when the new European privacy law, the General Data Protection Regulation (GDPR), was adopted on April 14, 2016. The general public, however, hardly seemed to take notice. Yet when the law finally went into effect on May 25, 2018 (after a preparation period of two years), it was hard to escape the many news stories and TV specials discussing its implications. Inboxes across the world filled up with emails from companies and organizations who updated their privacy policies to comply with the new law, or asked recipients to confirm long-forgotten subscriptions. Is this the watershed moment for privacy, when it finally moves from legal niche into mainstream public policy, leading to the widespread adoption of privacy-enhancing technologies in everything from Web servers to mobile devices to smart toys and smart homes?

As we laid out in this synthesis lecture, easy solutions for addressing privacy in mobile and pervasive computing might be hard to come by. Privacy is a complex topic with a rich history, a complex socio-political underpinning, and challenging interactions and dependencies between technical, legal, and organizational processes. While the GDPR has brought the concept of "privacy by design" into the spotlight, developing a systematic practice for integrating privacy measures into systems is still an ongoing challenge. What are the right privacy defaults? What is the absolute minimal data needed for a particular service? How can one limit the use of data without restricting future big data services? How does one make complex information and data flows transparent for users? How can one obtain consent from indviduals that is specific and freely given, without inundating users with prompts and messages? Or should we abandon the idea of "notice & choice" in a world full of mobile and pervasive computing? What is the right way to anonymize data? Is anonymization even possible in practical applications, given the ability to re-identify people by merging multiple innocuous datasets? How can we organize a fair marketplace around personal data, and what is the value of my data—today, and tomorrow? Who owns my data, if an "ownership" concept even makes sense for personal data?

The ability of mobile and pervasive computing systems to collect, analyze, and use personal data are ever increasing, with each new generation of technology being smaller, more power-efficient, and more ubiquitous. Despite the already substantial body of research in that area that we discussed here, ever more research and engineering challenges regarding privacy in mobile and pervasive computing continue to emerge. However, the welcome thrust from the policy side through the GDPR may help to further unify the often diverse research efforts in this space. An interdisciplinary approach, combining research in psychology, economics, law, social science, and computer science, stands the best chance to make progress in this complex field. Some of the key challenges that we see are the following.

- *Refining privacy primitives.* At the outset, research needs to continue investigating the fundamental principles of privacy-aware systems and privacy-enhancing technologies, with a particular focus on big data and anonymization.

- *Addressing system privacy.* The increasing interconnection of mobile and pervasive computing systems requires effective means of regulating access and use of both personal and anonymous (but potentially re-identifiable) data.

- *Supporting usable privacy.* Privacy solutions too often place a burden on users. How can privacy be understandable (and controllable) for end users, not just lawyers? How can legal requirements be reconciled with user experience requirements?

- *Personalizing privacy.* Will we be able to create systems that can adapt to individual privacy needs without being paternalistic? Can we find solutions that scale to millions yet provide the right support and effective assistance in managing privacy for an individual?

- *Establishing privacy engineering.* While early proposals for a privacy-aware design process exist and privacy engineering is developing as practice, we need to better understand which process to use when and how to tailor privacy solutions to the characteristics and requirements of specific applications and their context. There will certainly be no "one-size fits all" solution.

- *Improving privacy evaluation.* Understanding what users want and how well a particular solution is working are key factors for establishing a rigorous scientific approach for privacy.

We hope that this Synthesis Lecture provides a useful starting point for exploring these challenges.

Marc Langheinrich and Florian Schaub
Lugano and Ann Arbor, October 2018.

Bibliography

Live monitoring helps engine manufacturers track performance. *The Telegraph Online*, Nov. 2010. URL https://www.telegraph.co.uk/travel/travelnews/8111075/Live-moni toring-helps-engine-manufacturers-track-performance.html. 63

20 Minuten. Cumulus: Migros musste daten der polizei geben. *20 Minuten*, Aug. 2004. URL http://www.20min.ch/news/kreuz_und_quer/story/29046904. 2

G. D. Abowd. What next, ubicomp?: celebrating an intellectual disappearing act. In *ACM Conference on Ubiquitous Computing (UbiComp '12)*, page 31, New York, USA, 2012. ACM. ISBN 9781450312240. DOI: 10.1145/2370216.2370222. 58

G. D. Abowd and E. D. Mynatt. Charting past, present, and future research in ubiquitous computing. *ACM Transactions on Computer-Human Interaction (TOCHI)*, 7(1):29–58, 2000. DOI: 10.1145/344949.344988. 45, 51, 52, 53

M. Ackerman and T. Darrell. Privacy in context. *Human-Computer Interaction*, 16(2-4):167–176, 2001. DOI: 10.1207/s15327051hci16234_03. 81

M. S. Ackermann and S. D. Mainwaring. Privacy issues and human-computer interaction. In L. F. Cranor and S. Garfinkel, editors, *Security and Usability*, chapter 19, pages 381–400. O'Reilly, 2005. ISBN 0-596-00827-9. 62, 76

A. Acquisti and C. M. Fong. An experiment in hiring discrimination via online social networks. resreport, SSRN, 2014. URL http://ssrn.com/abstract=2031979. DOI: 10.2139/ssrn.2031979. 67

A. Acquisti and J. Grossklags. Privacy and rationality in individual decision making. *IEEE Security & Privacy*, 3(1):26–33, Jan. 2005. ISSN 1540-7993. DOI: 10.1109/msp.2005.22. 73

A. Acquisti, R. Gross, and F. Stutzman. Face recognition and privacy in the age of augmented reality. *Journal of Privacy and Confidentiality*, 6(2):1, 2014. URL http://repository.cmu.edu/jpc/vol6/iss2/1/. DOI: 10.29012/jpc.v6i2.638. 66

A. Acquisti, L. Brandimarte, and G. Loewenstein. Privacy and human behavior in the age of information. *Science*, 347(6221):509–514, Jan. 2015. ISSN 0036-8075, 1095-9203. DOI: 10.1126/science.aaa1465. 3, 71, 73, 78, 80

A. Acquisti, I. Adjerid, R. Balebako, L. Brandimarte, L. F. Cranor, S. Komanduri, P. G. Leon, N. Sadeh, F. Schaub, M. Sleeper, Y. Wang, and S. Wilson. Nudges for privacy and security: understanding and assisting users' choices online. *ACM Computing Surveys*, 50(3):1–41, Aug. 2017. ISSN 03600300. DOI: 10.1145/3054926. 71, 73, 74, 76, 80

F. Adib, Z. Kabelac, and D. Katabi. Multi-person localization via RF body reflections. In *Proceedings of the 12th USENIX Symposium on Networked Systems Design and Implementation (NSDI 15)*, pages 279–292, 2015. ISBN 9781931971096. 64

Y. Agarwal and M. Hall. ProtectMyPrivacy: detecting and mitigating privacy leaks on iOS devices using crowdsourcing. In *Proceeding of the 11th Annual International Conference on Mobile Systems, Applications, and Services*, MobiSys '13, pages 97–110, New York, NY, USA, 2013. ACM. ISBN 978-1-4503-1672-9. DOI: 10.1145/2462456.2464460. 77

I. Altman. *The Environment and Social Behavior: Privacy, Personal Space, Territory, Crowding*. Brooks/Cole Publishing company, Monterey, California, 1975. 39, 79

R. Anderson. *Security Engineering*. Wiley, 2nd edition, 2008. ISBN 978-0-470-06852-6. URL `http://www.cl.cam.ac.uk/~rja14/book.html`. 58

APEC. Apec privacy framework. Asia-Pacific Economic Cooperation, 2017. URL `https://www.apec.org/Publications/2017/08/APEC-Privacy-Framework-(2015)`. 21

Apple. iPhone XS A12 Bionic. `https://www.apple.com/iphone-xs/a12-bionic/`, 2018. 46

H. Arendt. *The Human Condition*. University of Chicago Press, Chicago, 1958. DOI: 10.2307/40097657. 34

Article 29 Data Protection Working Party. Advice paper on essential elements of a definition and a provision on profiling within the eu general data protection regulation. `http://ec.europa.eu/justice/data-protection/article-29/documentation/other-document/files/2013/20130513_advice-paper-on-profiling_en.pdf`, May 2013. 65

P. Ashley, S. Hada, G. Karjoth, C. Powers, and M. Schunter. Enterprise privacy authorization language (EPAL 1.2). W3C member submission, W3C, 2003. 72

L. Atzori, A. Iera, and G. Morabito. The Internet of Things: a survey. *Computer Networks*, 54 (15):2787–2805, 2010. DOI: 10.1016/j.comnet.2010.05.010. 49

E. Baard. Buying trouble – your grocery list could spark a terror probe. *The Village Voice*, July 2002. URL `http://www.villagevoice.com/issues/0230/baard.php`. 30

M. Backes and D. Markus. Enterprise privacy policies and languages. In *Digital Privacy: Theory, Technologies, and Practices*, chapter 7, pages 135–153. Auerbach Publications, 2008. DOI: 10.1201/9781420052183.ch7. 72

M. M. Baig and H. Gholamhosseini. Smart health monitoring systems: an overview of design and modeling. *Journal of Medical Systems*, 37(2):9898, Apr. 2013. ISSN 0148-5598. DOI: 10.1007/s10916-012-9898-z. 50

T. M. Banks. GDPR matchup: Canada's personal information protection and electronic documents act. IAPP Privacy Tracker, May 2, 2017, 2017. URL https://iapp.org/news/a/matchup-canadas-pipeda-and-the-gdpr/. 21

J. Bardram and A. Friday. Ubiquitous computing systems. In J. Krumm, editor, *Ubiquitous Computing Fundamentals*, chapter 2, pages 37–94. CRC Press, 2009. DOI: 10.1201/9781420093612.ch2. 52

J. P. Barlow. A declaration of the independence of cyberspace. EFF Website. https://www.eff.org/cyberspace-independence 11

D. Barrett. One surveillance camera for every 11 people in Britain, says CCTV survey. *The Telegraph*, 2013. http://www.telegraph.co.uk/technology/10172298/One-surveillance-camera-for-every-11-people-in-Britain-says-CCTV-survey.html. 21

R. F. Baumeister and M. R. Leary. The need to belong: desire for interpersonal attachments as a fundamental human motivation. *Psychological bulletin*, 117(3):497, 1995. DOI: 10.1037/0033-2909.117.3.497. 34

R. Beckwith. Designing for ubiquity: the perception of privacy. *IEEE Pervasive Computing*, 2 (2):40–46, Apr. 2003. ISSN 1536-1268. DOI: 10.1109/mprv.2003.1203752. 49

K. Benitez and B. Malin. Evaluating re-identification risks with respect to the HIPAA privacy rule. *Journal of the American Medical Informatics Association : JAMIA*, 17(2):169–77, Jan. 2010. ISSN 1527-974X. DOI: 10.1136/jamia.2009.000026. 41

J. Bentham. Panopticon. In M. Bozovic, editor, *The Panopticon Writings (1995)*. Verso, London, 1787. 40

B. Berendt, O. Günther, and S. Spiekermann. Privacy in e-commerce: stated preferences vs. actual behavior. *Communications of the ACM*, 48(4):101–106, Apr. 2005. ISSN 00010782. DOI: 10.1145/1053291.1053295. 22

H. Berghel. Identity theft, social security numbers, and the Web. *Communications of the ACM*, 43(2):17–21, Feb. 2000. ISSN 00010782. DOI: 10.1145/328236.328114. 66

K. Bernsmed, I. A. Tøndel, and A. A. Nyre. Design and implementation of a CBR-based privacy agent. In *Seventh International Conference on Availability, Reliability and Security (ARES '12)*, pages 317–326. IEEE, Aug. 2012. ISBN 978-1-4673-2244-7. DOI: 10.1109/ares.2012.60. 80

J. Bhatia, T. D. Breaux, J. R. Reidenberg, and T. B. Norton. A theory of vagueness and privacy risk perception. In *2016 IEEE 24th International Requirements Engineering Conference (RE)*, pages 26–35, Sept. 2016. DOI: 10.1109/re.2016.20. 75

M. Billinghurst and T. Starner. Wearable devices: new ways to manage information. *Computer*, 32(1):57–64, Jan. 1999. ISSN 0018-9162. 00272. DOI: 10.1109/2.738305. 46

I. Bilogrevic, K. Huguenin, B. Agir, M. Jadliwala, and J.-P. Hubaux. Adaptive information-sharing for privacy-aware mobile social networks. In *ACM international joint conference on Pervasive and ubiquitous computing (UbiComp '13)*, page 657, New York, USA, 2013. ACM. ISBN 9781450317702. DOI: 10.1145/2493432.2493510. 77

A. Bogle. Who owns music, video, e-books after you die? *Slate.com*, Aug. 2014. URL `https://www.slate.com/blogs/future_tense/2014/08/22/digital_assets_and_d eath_who_owns_music_video_e_books_after_you_die.html?via=gdpr-consent`. 60

J. Bohn, V. Coroama, M. Langheinrich, F. Mattern, and M. Rohs. Social, economic, and ethical implications of ambient intelligence and ubiquitous computing. In W. Weber, J. M. Rabaey, and E. Aarts, editors, *Ambient Intelligence*, volume 10, chapter 1, pages 5–29. Springer, 2005. DOI: 10.1007/3-540-27139-2_2. 62

D. Brin. *The Transparent Society*. Perseus Books, Reading MA, 1998. 30

H. P. Brougham. *Historical Sketches of Statesmen Who Flourished in the Time of George III*, volume 1. Lea & Blanchard, Philadephia, PA, USA, 1839. As quoted in Platt [1989]. 8, 31

A. Buchenscheit, B. Könings, A. Neubert, F. Schaub, M. Schneider, and F. Kargl. Privacy implications of presence sharing in mobile messaging applications. In *MUM 2014*, pages 20–29. ACM Press, 2014. ISBN 9781450333047. DOI: 10.1145/2677972.2677980. 67

H. Burkert. Privacy - data protection – a German/European perspective. In C. Engel and K. H. Keller, editors, *Governance of Global Networks in the Light of Differing Local Values*, Law and Economics of International Telecommunications 43, pages 43–69. Nomos, Baden-Baden, 2000. URL `http://www.coll.mpg.de/sites/www/files/text/burkert.pdf`. 43

L. A. Bygrave. *Data Privacy Law - An International Perspective*. Oxford University Press, Jan. 2014. ISBN 9780199675555. DOI: 10.1093/acprof:oso/9780199675555.001.0001. 43

R. Caceres and A. Friday. Ubicomp systems at 20: progress, opportunities, and challenges. *IEEE Pervasive Computing*, 11(1):14–21, Jan. 2012. ISSN 1536-1268. DOI: 10.1109/mprv.2011.85. 52

R. Carroll and S. Prickett, editors. *The Bible*. Oxford University Press, Oxford, UK, 2008. ISBN 978-0199535941. DOI: 10.1093/oseo/instance.00016818. 8, 9

J. Cas. Privacy in pervasive computing environments - a contradiction in terms? *IEEE Technology and Society Magazine*, 24(1):24–33, Jan. 2005. ISSN 0278-0097. DOI: 10.1109/mtas.2005.1407744. 62

F. H. Cate. The Limits of Notice and Choice. *IEEE Security & Privacy*, 8(2):59–62, 2010. ISSN 1540-7993. DOI: 10.1109/msp.2010.84. 73, 75, 76

F. H. Cate. *Privacy in the Information Age*. The Brookings Institution, Washington, D.C., USA, online edition, 1997. URL `brookings.nap.edu/books/0815713169/html`. 15, 37, 121

F. H. Cate. The failure of fair information practice principles. In J. K. Winn, editor, *Consumer Protection in the Age of the 'Information Economy'*, chapter 13, pages 341–378. Routledge, 2006. URL `https://ssrn.com/abstract=1156972`. 14

A. Cavoukian. *Privacy by Design ... Take the Challenge*. Information and Privacy Commissioner of Ontario, Canada, 2009. URL `http://privacybydesign.ca`. 87

R. N. Charette. Michigan's MiDAS unemployment system: algorithm alchemy created lead, not gold - IEEE spectrum. *IEEE Spectrum*, 18(3):6, 2018. URL `https://spectrum.ieee.org/riskfactor/computing/software/michigans-midas-unemployment-system-algorithm-alchemy-that-created-lead-not-gold`. 67, 83

D. Chen, S. P. Fraiberger, R. Moakler, and F. Provost. Enhancing transparency and control when drawing data-driven inferences about individuals. *Big Data*, 5(3):197–212, 2016. DOI: 10.1089/big.2017.0074. 84

S. Chitkara, N. Gothoskar, S. Harish, J. I. Hong, and Y. Agarwal. Does this app really need my location?: context-aware privacy management for smartphones. *Proc. ACM Interact. Mob. Wearable Ubiquitous Technol.*, 1(3):42:1–42:22, Sept. 2017. ISSN 2474-9567. DOI: 10.1145/3132029. 81

G. Chittaranjan, J. Blom, and D. Gatica-Perez. Who's who with Big-Five: analyzing and classifying personality traits with smartphones. In *2011 15th Annual International Symposium on Wearable Computers*, pages 29–36. IEEE, jun 2011. ISBN 978-1-4577-0774-2. DOI: 10.1109/iswc.2011.29. 66

R. Clarke. Beyond the OECD guidelines: privacy protection for the 21st century. `http://www.rogerclarke.com/DV/PP21C.html`, Jan. 2000. 11, 14

R. Clarke. What's 'privacy'? http://www.rogerclarke.com/DV/Privacy.html, Aug. 2006. DOI: 10.1163/9789004192195_004. 13, 14, 26, 43

R. Clarke. Privacy impact assessment: its origins and development. *Computer Law & Security Review*, 25(2):123–135, Jan. 2009. ISSN 0267-3649. DOI: 10.1016/j.clsr.2009.02.002. 88

P. Cochrane. Head to head. *Sovereign Magazine*, pages 56–57, 2000. URL http://www.cochrane.org.uk/opinion/papers/prof.htm. 24

J. E. Cohen. Examined lives: Informational privacy and the subject as object. *Stanford Law Review*, 52:1373–1437, May 2000. URL http://www.law.georgetown.edu/faculty/jec/examined.pdf. As cited in Solove and Rotenberg [2003]. DOI: 10.2307/1229517. 28

G. Cohn, S. Gupta, J. Froehlich, E. Larson, and S. N. Patel. GasSense: appliance-level, single-point sensing of gas activity in the home. In P. Floréen, A. Krüger, and M. Spasojevic, editors, *Pervasive Computing. Pervasive 2010.*, pages 265–282, Berlin, Heidelberg, 2010. Springer. DOI: 10.1007/978-3-642-12654-3_16. 64

A. Compton. British tourists detained, deported for tweeting "destroy America", Jan. 2012. URL http://www.huffingtonpost.com/2012/01/30/british-tourists-deported-for-tweeting_n_1242073.html. 23

J. Cornwell, I. Fette, G. Hsieh, M. Prabaker, J. Rao, K. Tang, K. Vaniea, L. Bauer, L. F. Cranor, J. Hong, B. McLaren, M. Reiter, and N. Sadeh. User-controllable security and privacy for pervasive computing. In *Eighth IEEE Workshop on Mobile Computing Systems and Applications (HotMobile '07)*, pages 14–19. IEEE, Mar. 2007. ISBN 0-7695-3001-X. DOI: 10.1109/wmcsa.2007.4389552. 77

Council of Europe. Convention for the protection of human rights and fundamental freedoms. CETS 005, Nov. 1950. URL conventions.coe.int/Treaty/en/Treaties/Html/005.htm. 10

Council of Europe. Resolution (73) 22 on the protection of the privacy of individuals vis-à-vis electronic data banks in the private sector, 1973. URL http://www.coe.int/T/E/Legal_affairs/Legal_co-operation/Data_protection/Documents/International_legal_instruments/Resolution%20(73)%2022.asp. 18

Council of Europe. Resolution (74) 29 on the protection of the privacy of individuals vis-à-vis electronic data banks in the public sector, 1974. URL http://www.coe.int/T/E/Legal_affairs/Legal_co-operation/Data_protection/Documents/International_legal_instruments/Resolution%20(74)%2029.asp. 18

Council of Europe. Convention for the protection of individuals with regard to automatic processing of personal data. CETS 108, Jan. 1981. URL conventions.coe.int/Treaty/en/Treaties/Html/108.htm. 18, 20

Council of Europe. Convention 108+ – Modernised convention for the protection of individuals with regard to automatic processing of personal data. CETS 108+, June 2018. URL `http://rm.coe.int/convention-108-convention-for-the-protection-of-in dividuals-with-regar/16808b36f1`. 20

L. F. Cranor. Necessary but not sufficient: standardized mechanisms for privacy notice and choice. *Journal of Telecommunications and High Technology Law*, 10(2), 2012. URL `http://jthtl.org/content/articles/V10I2/JTHTLv10i2_Cranor.PDF`. 72, 76

J. Cranshaw, J. Mugan, and N. Sadeh. User-controllable learning of location privacy policies with gaussian mixture models. In *25th AAAI Conference on Artifical Intelligence*. AAAI, 2011. URL `https://www.aaai.org/ocs/index.php/AAAI/AAAI11/paper/viewPaper/3785`. 77

K. Crawford and J. Schultz. Big data and due process: toward a framework to redress predictive privacy harms. *Boston College Law Review*, 55(93), Oct. 2014. URL `https://papers.ssrn.com/sol3/papers.cfm?abstract_id=2325784http://la wdigitalcommons.bc.edu/bclr/vol55/iss1/4/`. 84

M.-F. Cuéllar. Cyberdelegation and the administrative state. Stanford public law working paper, Stanford University, Oct. 2016. DOI: 10.1017/9781316671641.006. 85

G. Danezis, J. Domingo-Ferrer, M. Hansen, J.-H. Hoepman, D. L. Metayer, R. Tirtea, and S. Schiffner. Privacy and data protection by design - from policy to engineering. Technical Report December, ENISA - European Union Agency for Network and Information Security, 2015. DOI: 10.2824/38623. 87, 88

A. Das, M. Degeling, D. Smullen, and N. Sadeh. Personalized privacy assistants for the Internet of Things. *IEEE Pervasive Computing*, 2018. DOI: 10.1109/mprv.2018.03367733. 77, 80, 81

T. Das, P. Mohan, V. N. Padmanabhan, R. Ramjee, and A. Sharma. PRISM: platform for remote sensing using smartphones. In *Proceedings of the 8th international conference on Mobile systems, applications, and services - MobiSys '10*, page 63, New York, USA, 2010. ACM Press. ISBN 9781605589855. DOI: 10.1145/1814433.1814442. 53

N. Davies, M. Langheinrich, S. Clinch, I. Elhart, A. Friday, T. Kubitza, and B. Surajbali. Personalisation and privacy in future pervasive display networks. In *Proceedings of the SIGCHI Conference on Human Factors in Computing Systems*, CHI '14, pages 2357–2366, New York, USA, 2014. ACM. ISBN 978-1-4503-2473-1. 00000. DOI: 10.1145/2556288.2557287. 80

P. De Hert and V. Papakonstantinou. The proposed data protection Regulation replacing Directive 95/46/EC: a sound system for the protection of individuals. *Computer Law & Security Review*, 28(2):130–142, Apr. 2012. ISSN 0267-3649. DOI: 10.1016/j.clsr.2012.01.011. 19

P. De Hert and V. Papakonstantinou. The new general data protection regulation: still a sound system for the protection of individuals? *Computer Law & Security Review*, 32(2):179–194, Apr. 2016. ISSN 0267-3649. DOI: 10.1016/j.clsr.2016.02.006. 19, 43

P. De Hert, S. Gutwirth, A. Moscibroda, D. Wright, and G. González Fuster. Legal safeguards for privacy and data protection in ambient intelligence. *Personal and Ubiquitous Computing*, 13(6):435–444, Oct. 2009. ISSN 1617-4909. DOI: 10.1007/s00779-008-0211-6. 62

Der Spiegel, 2004. Innere sicherheit: Totes pferd. *Der Spiegel*, (11):48, Mar. 2004. URL http://www.spiegel.de/spiegel/inhalt/0,1518,ausg-1395,00.html. 26

N. Dhingra, Z. Gorn, A. Kener, and J. Dana. The default pull: an experimental demonstration of subtle default effects on preferences. *Judgment and Decision Making*, 7(1):69–76, 2012. ISSN 1930-2975. 71

N. Diakopoulos. Accountability in algorithmic decision making. *Communications of the ACM*, 59(2):56–62, Jan. 2016. ISSN 00010782. DOI: 10.1145/2844110. 83

A. S. Douglas. The U.K. privacy white paper 1975. In *Proceedings of the June 7-10, 1976, National Computer Conference and Exposition*, AFIPS '76, pages 33–38, New York, USA, 1976. ACM. DOI: 10.1145/1499799.1499806. 58

P. Dourish. What we talk about when we talk about context. *Personal and Ubiquitous Computing*, 8(1):19–30, Feb. 2004. ISSN 1617-4909. DOI: 10.1007/s00779-003-0253-8. 52

B. Dumas, D. Lalanne, and S. Oviatt. Multimodal interfaces: a survey of principles, models and frameworks. *Human Machine Interaction*, pages 3–26, 2009. DOI: 10.1007/978-3-642-00437-7_1. 51

M. R. Ebling and M. Baker. Pervasive tabs, pads, and boards: are we there yet? *IEEE Pervasive Computing*, 11(1):42–51, Jan. 2012. ISSN 1536-1268. DOI: 10.1109/mprv.2011.80. 47

A. Etzioni. *The Limits of Privacy*. Basic Books, New York, USA, 1999. DOI: 10.2307/2654355. 28

European Commission. Communication from the Commission to the European Parliament and the Council: Exchanging and protecting personal data in a globalised world. Press Release, Jan. 2017. URL http://ec.europa.eu/newsroom/document.cfm?doc_id=41157. 20

European Parliament. Charter of fundamental rights of the European Union. *Official Journal of the European Communities*, 55(C 326):391–407, Oct. 2000. ISSN 1977-091X. DOI: 10.1515/9783110971965. 10

European Parliament and Council. Directive 95/46/EC on the protection of individuals with regard to the processing of personal data and on the free movement of such data. *Official Journal of the European Communities*, 38(L 281):31–50, Nov. 1995. URL `https://eur-le x.europa.eu/eli/dir/1995/46/oj`. 18, 72, 85

European Parliament and Council. Directive 2002/58/EC concerning the processing of personal data and the protection of privacy in the electronic communications sector (directive on privacy and electronic communications). *Official Journal of the European Communities*, 45 (L 201):37–47, July 2002. ISSN 0378-6978. URL `https://eur-lex.europa.eu/legal-content/EN/TXT/?uri=CELEX:32002L0058`. 26

European Parliament and Council. Regulation (EU) 2016/679 of the European parliament and of the council of 27 april 2016 on the protection of natural persons with regard to the processing of personal data and on the free movement of such data, and repealing directive 95/46/EC (general data protection regulation). *Official Journal of the European Communities*, 59(L 119):1–88, May 2016. ISSN 1977-0677. URL `https://eur-lex.europa.eu/eli/ reg/2016/679/oj`. 13, 43, 71, 72, 76, 81, 84, 85, 87, 89

Facebook, Inc. Did you know that… 4.75 billion pieces of content are shared daily?, 2013. URL `https://www.facebook.com/FacebookSingapore/posts/563468333703369`. 22

Facebook, Inc. Quarterly earning, 2018. URL `https://investor.fb.com/financials/?se ction=quarterlyearnings`. 22

Federal Trade Commission. Privacy online: A report to congress. Staff report, Federal Trade Commission, Federal Trade Commission, Washington, DC, June 1998. URL `https://www.ftc.gov/sites/default/files/documents/reports/privacy-on line-report-congress/priv-23a.pdf`. 73

Federal Trade Commission. Protecting consumer privacy in an era of rapid change. Staff report, Federal Trade Commission, Federal Trade Commission, Washington, DC, Mar. 2012. URL `https://www.ftc.gov/sites/default/files/documents/reports/federal-trade-commission-report-protecting-consumer-privacy-era-rapid-change-recommendations/120326privacyreport.pdf`. 76

Federal Trade Commission. Internet of Things: Privacy and security in a connected world. Staff report, Federal Trade Commission, Federal Trade Commission, Washington, DC, Jan. 2015. URL `http://www.ftc.gov/system/files/documents/reports/federal-trade-commission-staff-report-november-2013-workshop-entitled-internet-things-privacy/150127iotrpt.pdf`. 64, 76

A. Ferscha. 20 years past Weiser - what next? *IEEE Pervasive Computing*, 11(1):52–61, 2012. ISSN 1536-1268. DOI: 10.1109/mprv.2011.78. 45, 53, 54

D. H. Flaherty. *Protecting Privacy in Surveillance Societies: The Federal Republic of Germany, Sweden, France, Canada, and the United States.* University of North Carolina Press, Chapel Hill, NC, USA, 1989. 10

E. Fleisch, F. Mattern, and S. Billinger. Betriebswirtschaftliche applikationen des ubiquitous computing - beispiele, bausteine und nutzenpotentiale. In H. Sauerburger, editor, *Ubiquitous Computing*, number 229 in HMD – Praxis der Wirtschaftsinformatik, pages 5–15. dpunkt.verlag, Feb. 2003. ISBN 3-89864-200-3. DOI: 10.1007/978-3-642-55550-3_6. 63

J. Flinn. Cyber Foraging: Bridging Mobile and Cloud Computing. *Synthesis Lectures on Mobile and Pervasive Computing*, 7(2):1–103, Sept. 2012. ISSN 1933-9011. DOI: 10.2200/s00447ed1v01y201209mpc010. 47

M. Friedewald, O. D. Costa, Y. Punie, P. Alahuhta, and S. Heinonen. Perspectives of ambient intelligence in the home environment. *Telematics and Informatics*, 22(3):221–238, Aug. 2005. ISSN 07365853. DOI: 10.1016/j.tele.2004.11.001. 50

J. E. Froehlich, E. Larson, T. Campbell, C. Haggerty, J. Fogarty, and S. N. Patel. HydroSense: infrastructure-mediated single-point sensing of whole-home water activity. In H.-W. Gellersen, S. Consolvo, and S. Helal, editors, *Proceedings of the 11th international conference on Ubiquitous computing - Ubicomp '09*, page 235, New York, USA, 2009. ACM Press. ISBN 9781605584317. DOI: 10.1145/1620545.1620581. 64

P. Fule and J. F. Roddick. Detecting privacy and ethical sensitivity in data mining results, 2004. URL https://dl.acm.org/citation.cfm?id=979942. 85

R. Ganti, F. Ye, and H. Lei. Mobile crowdsensing: current state and future challenges. *IEEE Communications Magazine*, 49(11):32–39, Nov. 2011. ISSN 0163-6804. DOI: 10.1109/mcom.2011.6069707. 53

R. Gavison. Privacy and the limits of the law. *Yale Law Journal*, 89:421–471, 1980. DOI: 10.2307/795891. 104

R. Gavison. Privacy and the limits of the law. In F. D. Schoeman, editor, *Philosophical Dimensions of Privacy: An Anthology*, pages 346–402. Cambridge University Press, Cambridge, UK, 1984. Originally published in Gavison [1980] and reprinted as Gavison [1995]. DOI: 10.1017/cbo9780511625138.017. 36

R. Gavison. Privacy and the limits of the law. In D. G. Johnson and H. Nissenbaum, Eds., *Computers, Ethics, and Social Values*, pages 332–351, Prentice Hall, Inc., 1995. DOI: 10.1017/cbo9780511625138.017. 104

R. S. Gerstein. Intimacy and privacy. *Ethics*, 89(1):76–81, Oct. 1978. URL http://www.jsto r.org/stable/2380133. DOI: 10.1086/292105. 34

J. Ginsberg, M. H. Mohebbi, R. S. Patel, L. Brammer, M. S. Smolinski, and L. Brilliant. Detecting influenza epidemics using search engine query data. *Nature*, 457(7232):1012–1014, 2009. DOI: 10.1038/nature07634. 62

I. Glass. This American life: right to remain silent, 2010. URL https://www.thisamerican life.org/414/transcript. 23

P. Golle and K. Partridge. On the anonymity of home/work location pairs. In H. Tokuda, M. Beigl, A. Friday, A. J. B. Brush, and Y. Tobe, editors, *7th International Conference on Pervasive Computing (Pervasive '09)*, pages 390–397. Springer, 2009. ISBN 978-3-642-01515-1. DOI: 10.1007/978-3-642-01516-8_26. 41

Google. Google now. Website, 2014. URL https://www.google.com/landing/now/. 48, 61

Google. Material design guidelines: Permissions. Website, Sept. 2017. URL https://materi al.io/design/platform-guidance/android-permissions.html. 74, 75

K. Gormley. 100 years of privacy. *Wisconsin Law Review*, pages 1335–1442, 1992. ISSN 09547762. URL http://heinonlinebackup.com/hol-cgi-bin/get_pdf.cgi?handle =hein.journals/wlr1992&section=57. 15, 43

G. Greenleaf. *Asian Data Privacy Laws*. Oxford University Press, oct 2014. ISBN 9780199679669. DOI: 10.1093/acprof:oso/9780199679669.001.0001. 43

G. Greenleaf. Global data privacy laws 2017: 120 national data privacy laws, including Indonesia and Turkey. *Privacy Laws & Business International Report*, 2017(145):10–13, Jan. 2017. URL https://papers.ssrn.com/sol3/papers.cfm?abstract_id=2993035. 20, 24, 43

G. Greenleaf. Convention 108+ and the data protection framework of the EU (speaking notes for conference presentation at 'Convention 108+ tomorrow's common ground for protection'). June 2018. URL https://papers.ssrn.com/sol3/papers.cfm?abstract_id=3202606. 20

U. Greveler, B. Justus, and D. Loehr. Multimedia content identification through smart meter power usage profiles. In *Computers, Privacy and Data Protection*, 2012. 61, 64

M. Gruteser and B. Hoh. On the anonymity of periodic location samples. In *Second international conference on Security in Pervasive Computing (SPC '05)*. Springer, 2005. DOI: 10.1007/978-3-540-32004-3_19. 41

Guardian. The card up their sleeve. *The Guardian*, July 19, 2003. URL `http://www.guardian.co.uk/weekend/story/0,3605,999866,00.html`. 30

S. Gupta, M. S. Reynolds, and S. N. Patel. ElectriSense: single-point sensing using emi for electrical event detection and classification in the home. In K. N. Truong, P. A. Nixon, J. E. Bardram, and M. Langheinrich, editors, *Proceedings of the 12th ACM international conference on Ubiquitous computing - Ubicomp '10*, pages 139–148, New York, USA, 2010. ACM Press. ISBN 9781605588438. DOI: 10.1145/1864349.1864375. 64

S. Gürses and J. M. del Alamo. Privacy engineering: shaping an emerging field of research and practice. *IEEE Security & Privacy*, 14(2):40–46, Mar. 2016. ISSN 1540-7993. DOI: 10.1109/msp.2016.37. 87

S. Gürses, C. Troncoso, and C. Diaz. Engineering privacy by design reloaded. *Amsterdam Privacy Conference 2015 (APC15)*, (610613):1–21, 2015. ISSN 1095-9203. 86

R. Hartley-Parkinson. 'I'm going to destroy America and dig up Marilyn Monroe': British pair arrested in U.S. on terror charges over Twitter jokes, jan 2012. URL `http://www.dailymail.co.uk/news/article-2093796/Emily-Bunting-Leigh-Van-Bryan-UK-tourists-arrested-destroy-America-Twitter-jokes.html`. 23

K. Hawkey and K. M. Inkpen. Keeping up appearances: understanding the dimensions of incidental information privacy. In *SIGCHI conference on Human Factors in computing systems (CHI '06)*, page 821, New York, USA, 2006. ACM. ISBN 1595933727. DOI: 10.1145/1124772.1124893. 41

A. Hern. Samsung rejects concern over 'orwellian' privacy policy. The Guardian, Feb. 9 2015. URL `http://www.theguardian.com/technology/2015/feb/09/samsung-rejects-concern-over-orwellian-privacy-policy`. 3

A. Hern. Technology: most GDPR emails unnecessary and some illegal, say experts, May 2018. URL `https://www.theguardian.com/technology/2018/may/21/gdpr-emails-mostly-unnecessary-and-in-some-cases-illegal-say-experts`. 20

HEW Advisory Committee. Records, computers and the rights of citizens – report of the secretary's advisory committee on automated personal data systems, records, computers and the rights of citizens. Technical report, U.S. Department of Health, Education, and Welfare (HEW), 1973. URL `http://aspe.hhs.gov/datacncl/1973privacy/tocprefacemembers.htmhttps://epic.org/privacy/hew1973report/`. 11, 72, 73

M. Hildebrandt. Profiling and the identity of the European citizen. In *Profiling the European Citizen*, pages 303–343. Springer Netherlands, Dordrecht, 2008. DOI: 10.1007/978-1-4020-6914-7_15. 83

K. Hill. Officemax blames data broker for 'daughter killed in car crash'. *Forbes Online*, Jan. 2014. URL `https://www.forbes.com/sites/kashmirhill/2014/01/22/officemax-blam` `es-data-broker-for-daughter-killed-in-car-crash-letter/{#}50f16b0f76cf.` 66

J.-H. Hoepman. Privacy design strategies. *arXiv preprint arXiv:1210.6621*, 9:12, 2012. DOI: 10.1007/978-3-642-55415-5_38. 87

M. N. Husen and S. Lee. Indoor human localization with orientation using WiFi fingerprinting. In *Proceedings of the 8th International Conference on Ubiquitous Information Management and Communication - ICUIMC '14*, pages 1–6, New York, USA, 2014. ACM Press. ISBN 9781450326445. DOI: 10.1145/2557977.2557980. 64

IBM Corp. IBM builds a smarter planet. `https://www.ibm.com/smarterplanet/us/en/`, 2008. 50

IBM Global Services. IBM multi-national privacy survey. Consumer Report 938568, Harris Interactive, New York, USA, Oct. 1999. URL `http://web.asc.upenn.edu/usr/ogandy` `/ibm_privacy_survey_oct991.pdf.` 22

ICO. Privacy impact assessment handbook. Technical report, Information Commissioner's Office (ICO), Wilmslow, Cheshire, 2007. URL `http://ico.org.uk/for_organisations` `/data_protection/topic_guides/privacy_impact_assessment.` 88

ICO. Guide to the General Data Protection Regulation (GDPR). Technical report, Information Commissioner's Office (ICO), Wilmslow, Cheshire, Sept. 2018. URL `http://www.ico.org.uk/media/for-organisations/documents/1595/pi` `a-code-of-practice.pdf.` 88

IFTTT. IFTTT Website, Nov. 2014. URL `https://ifttt.com/.` 47

International Data Corporation (IDC). New IDC survey finds widespread privacy concerns among U.S. consumers (IDC US42238617), Jan. 2017. URL `https://www.idc.com/getd` `oc.jsp?containerId=prUS42253017.` [Online; posted 24-January-2017]. 22

Internetlivestats.com. Google search statistics, 2018. URL `http://www.internetlivestat` `s.com/google-search-statistics/.` 22

H. Ishii and B. Ullmer. Tangible bits: towards seamless interfaces between people, bits and atoms. In *SIGCHI conference on Human factors in computing systems (CHI '97)*, pages 234–241, New York, USA, 1997. ACM. ISBN 0897918029. DOI: 10.1145/258549.258715. 51

J. James. Data never sleeps 5.0, 2017. URL `https://www.domo.com/learn/data-never-` `sleeps-5.` 22

E. J. Janger and P. M. Schwartz. The gramm-leach-bliley act, information privacy, and the limits of default rules. *Minnesota Law Review*, 1230(89):1219, 2002. ISSN 1556-5068. DOI: 10.2139/ssrn.319144. 71

C. Jernigan and B. F. T. Mistree. Gaydar: Facebook friendships expose sexual orientation. *First Monday*, 14(10), 2009. URL http://firstmonday.org/article/view/2611/2302. DOI: 10.5210/fm.v14i10.2611. 67, 84

B. Jones, R. Sodhi, M. Murdock, R. Mehra, H. Benko, A. Wilson, E. Ofek, B. MacIntyre, N. Raghuvanshi, and L. Shapira. RoomAlive: magical experiences enabled by scalable, adaptive projector-camera units. In *Proceedings of the 27th Annual ACM Symposium on User Interface Software and Technology*, UIST '14, pages 637–644, New York, USA, 2014. ACM. ISBN 978-1-4503-3069-5. DOI: 10.1145/2642918.2647383. 52

R. Jones. Consumer rights: why the GDPR email deluge, and can i ignore it?, May 2018. URL https://www.theguardian.com/money/2018/may/12/why-the-gdpr-email-deluge-and-can-i-ignore-it. 20

W. Jones. Building safer cars. *IEEE Spectrum*, 39(1):82–85, 2002. ISSN 00189235. DOI: 10.1109/6.975028. 50

S. Joyee De and D. Le Métayer. *Privacy Risk Analysis*, volume 81 of *Synthesis Lectures on Information Security, Privacy, and Trust*. Morgan & Claypool, 2016. ISBN 9781627059879. DOI: 10.2200/S00724ED1V01Y201607SPT017. 25

F. Kargl. *Inter-Vehicular Communication*. Habilitation thesis, Ulm University, Ulm, Dec. 2008. 53

D. Katz. Top 10 fitness APIs: Apple health, Fitbit and Nike, 2015. URL https://www.programmableweb.com/news/top-10-fitness-apis-apple-health-fitbit-and-nike/analysis/2015/04/17. 63

R. Kazman, M. Klein, M. Barbacci, T. Longstaff, H. Lipson, and J. Carriere. The architecture tradeoff analysis method. Technical report, Software Engineering Institute (SEI), Pittsburgh, PA, USA, 1998. URL http://ieeexplore.ieee.org/abstract/document/706657/. DOI: 10.21236/ada350761. 89

P. G. Kelley, P. Hankes Drielsma, N. Sadeh, and L. F. Cranor. User-controllable learning of security and privacy policies. In *1st ACM workshop on AISec (AISec '08)*, page 11, New York, USA, 2008. ACM. ISBN 9781605582917. DOI: 10.1145/1456377.1456380. 76, 77

W. Knight. The dark secret at the heart of AI. *MIT Technology Review*, 17(5):54, 2017. URL https://www.technologyreview.com/s/604087/the-dark-secret-at-the-heart-of-ai/. 84

B. P. Knijnenburg and A. Kobsa. Helping users with information disclosure decisions: potential for adaptation. In *International conference on Intelligent user interfaces (IUI '13)*, page 407, New York, USA, 2013. ACM. ISBN 9781450319652. DOI: 10.1145/2449396.2449448. 76, 77, 80

J. Koesters. Smart shelves will help stock supermarkets of the future, 2018. URL http://www.digitalistmag.com/iot/2018/02/28/smart-shelves-will-help-stock-supermarkets-of-future-05922036. 63

B. Könings and F. Schaub. Territorial privacy in ubiquitous computing. In *Eighth International Conference on Wireless On-Demand Network Systems and Services (WONS '11)*, pages 104–108, Bardoneccia, Jan. 2011. IEEE. ISBN 978-1-61284-189-2. DOI: 10.1109/wons.2011.5720177. 81

B. Könings, C. Bachmaier, F. Schaub, and M. Weber. Device names in the wild: investigating privacy risks of zero configuration networking. In *Workshop on Privacy and Security for Moving Objects (PriSMO), IEEE 14th International Conference on Mobile Data Management (MDM '13)*, pages 51–56. IEEE, June 2013. ISBN 978-0-7695-4973-6. DOI: 10.1109/mdm.2013.65. 67

B. Könings, S. Thoma, F. Schaub, and M. Weber. PriPref broadcaster: enabling users to broadcast privacy preferences in their physical proximity. In *MUM '14 Proceedings of the 13th International Conference on Mobile and Ubiquitous Multimedia*, pages 133–142. ACM Press, 2014. ISBN 9781450333047. DOI: 10.1145/2677972.2677978. 80

B. Könings, F. Schaub, and M. Weber. *Privacy and Trust in Ambient Intelligent Environments*, pages 133–164. Springer International Publishing, Cham, 2016. ISBN 978-3-319-23452-6. DOI: 10.1007/978-3-319-23452-6_4. 80

B.-J. Koops. The (in)flexibility of techno-regulation and the case of purpose-binding. *Legisprudence*, 5(2):171–194, Oct. 2011. ISSN 1752-1467. DOI: 10.5235/175214611797885701. 72

I. Kottasová. These companies are getting killed by GDPR, May 2018. URL https://money.cnn.com/2018/05/11/technology/gdpr-tech-companies-losers/index.html. 20

KPMG. Companies that fail to see privacy as a business priority risk crossing the "creepy line", Nov. 2016. URL https://home.kpmg.com/sg/en/home/media/press-releases/2016/11/companies-that-fail-to-see-privacy-as-a-business-priority-risk-crossing-the-creepy-line.html. [Online; posted 6-November-2016]. 22

I. Kroener and D. Wright. A strategy for operationalizing privacy by design. *Information Society*, 30(5), 2014. ISSN 10876537. DOI: 10.1080/01972243.2014.944730. 88, 89

P. Kumaraguru and L. F. Cranor. Privacy indexes : a survey of westin's studies. Technical Report CMU-ISRI-5-138, Carnegie Mellon University, Institute for Software Research, 2005. URL `http://repository.cmu.edu/isr/856/`. 22

O. Kwon. A pervasive P3P-based negotiation mechanism for privacy-aware pervasive e-commerce. *Decision Support Systems*, 50(1):213–221, Dec. 2010. ISSN 01679236. DOI: 10.1016/j.dss.2010.08.002. 80

S. Lahlou, M. Langheinrich, and C. Röcker. Privacy and trust issues with invisible computers. *Communications of the ACM*, 48(3):59–60, Mar. 2005. ISSN 00010782. DOI: 10.1145/1047671.1047705. 4

S. Landau. Making Sense from Snowden: what's significant in the NSA surveillance revelations. *IEEE Security & Privacy*, 11(4):54–63, July 2013. ISSN 1540-7993. DOI: 10.1109/msp.2013.90. 60, 79

S. Landau. Highlights from Making Sense of Snowden, Part II: what's significant in the NSA revelations. *IEEE Security & Privacy*, 12(1):62–64, Jan. 2014. ISSN 1540-7993. DOI: 10.1109/msp.2013.161. 60, 79

N. D. Lane, Y. Chon, L. Zhou, Y. Zhang, F. Li, D. Kim, G. Ding, F. Zhao, and H. Cha. Piggyback CrowdSensing (PCS): energy efficient crowdsourcing of mobile sensor data by exploiting smartphone app opportunities. *SenSys '13 Proceedings of the 11th ACM Conference on Embedded Networked Sensor Systems*, New York, USA. ACM. DOI: 10.1145/2517351.2517372. 53

M. Langheinrich. A privacy awareness system for ubiquitous computing environments. In *4th International Conference on Ubiquitous Computing (UbiComp '02)*, pages 237–245. Springer, 2002a. ISBN 9783540442677. DOI: 10.1007/3-540-45809-3_19. 80, 81

M. Langheinrich. Privacy invasions in ubiquitous computing. UbiComp '02 Workshop on Socially-Informed Design of Privacy-Enhancing Solutions, Sept. 2002b. URL `http://www.inf.usi.ch/faculty/langheinrich/articles/papers/langhein2002-ucpws.pdf`. DOI: 10.1201/9781420093612.ch3. 62

M. Langheinrich. Privacy in ubiquitous computing. In J. Krumm, editor, *Ubiqutious Computing Fundamentals*, chapter 3, pages 95–160. CRC Press, 2009. DOI: 10.1201/9781420093612. 52, 57, 58, 62

S. Laughlin. Ces 2017: The year of voice. J. Walter Thompson Intelligence, 2017. URL `https://www.jwtintelligence.com/2017/01/ces-2017-year-voice/`. 3

C. Laurant, editor. *Privacy and Human Rights 2003*. EPIC and Privacy International, London, UK, 2003. ISBN 1-893044-18-1. URL `http://www.privacyinternational.org/survey/phr2003/`. 8, 32

S. Lederer, J. I. Hong, A. K. Dey, and J. A. Landay. Five pitfalls in the design of privacy. In L. F. Cranor and S. Garfinkel, editors, *Security and Usability*, chapter 21, pages 421–446. O'Reilly, 2005. ISBN 0-596-00827-9. 62

T. B. Lee. Facebook's Cambridge Analytica scandal, explained. Ars Technica Website, Mar. 2018. URL `https://arstechnica.com/tech-policy/2018/03/facebooks-cambridge-analytica-scandal-explained/`. 29

Y.-D. Lee and W.-Y. Chung. Wireless sensor network based wearable smart shirt for ubiquitous health and activity monitoring. *Sensors and Actuators B: Chemical*, 140(2):390–395, July 2009. ISSN 0925-4005. DOI: 10.1016/j.snb.2009.04.040. 50

J. T. Lehikoinen, J. Lehikoinen, and P. Huuskonen. Understanding privacy regulation in ubicomp interactions. *Personal and Ubiquitous Computing*, 12(8):543–553, Mar. 2008. ISSN 1617-4909. DOI: 10.1007/s00779-007-0163-2. 79

L. Lessig. *Code and Other Laws of Cyberspace*. Basic Books, New York, USA, 1999. DOI: 10.1016/s0740-624x(00)00068-x. 25, 29, 30, 31, 43

G. K. Levinger and H. L. Raush. *Close Relationships: Perspectives on the Meaning of Intimacy*. University of Massachusetts Press, 1977. DOI: 10.2307/351497. 34

J. Leyden. FBI apology for Madrid bomb fingerprint fiasco. *The Register*, May 26, 2004. URL `http://www.theregister.co.uk/2004/05/26/fbi_madrid_blunder/`. 30

Y. Li, F. Chen, T. J.-J. Li, Y. Guo, G. Huang, M. Fredrikson, Y. Agarwal, and J. I. Hong. PrivacyStreams: enabling transparency in personal data processing for mobile apps. *Proc. ACM Interact. Mob. Wearable Ubiquitous Technol.*, 1(3):76:1–76:26, Sept. 2017. ISSN 2474-9567. DOI: 10.1145/3130941. 81

J. Lin, S. Amini, J. I. Hong, N. Sadeh, J. Lindqvist, and J. Zhang. Expectation and purpose: understanding users' mental models of mobile app privacy through crowdsourcing. In *ACM Conference on Ubiquitous Computing (Ubicomp '12)*. ACM, 2012. DOI: 10.1145/2370216.2370290. 77

J. Lin, B. Liu, N. Sadeh, and J. I. Hong. Modeling users' mobile app privacy preferences: restoring usability in a sea of permission settings. In *10th Symposium On Usable Privacy and Security (SOUPS 2014)*, pages 199–212, Menlo Park, CA, 2014. USENIX Association. ISBN 978-1-931971-13-3. URL `https://www.usenix.org/conference/soups2014/proceedings/presentation/lin`. 77, 80

A. R. Lingley, M. Ali, Y. Liao, R. Mirjalili, M. Klonner, M. Sopanen, S. Suihkonen, T. Shen, B. P. Otis, H. Lipsanen, and B. A. Parviz. A single-pixel wireless contact lens display. *Journal of Micromechanics and Microengineering*, 21(12):125014, Dec. 2011. ISSN 0960-1317. DOI: 10.1088/0960-1317/21/12/125014. 50

B. Liu, M. S. Andersen, F. Schaub, H. Almuhimedi, S. A. Zhang, N. Sadeh, Y. Agarwal, and A. Acquisti. Follow my recommendations: a personalized privacy assistant for mobile app permissions. In *Twelfth Symposium on Usable Privacy and Security (SOUPS 2016)*, pages 27–41, Denver, CO, 2016. USENIX Association. ISBN 978-1-931971-31-7. URL `https://ww w.usenix.org/conference/soups2016/technical-sessions/presentation/liu`. 76, 77, 80

S. Lobo. Datenschutzgrundverordnung (DSGVO): Wer macht mir die geileren vorschriften? *Der Spiegel Online*, May 2018. URL `http://www.spiegel.de/netzwelt/web/datens chutzgrundverordnung-dsgvo-wer-macht-mir-die-geileren-vorschriften-a-1206979.html`. 20

C. R. Long and J. R. Averill. Solitude: An exploration of benefits of being alone. *Journal for the Theory of Social Behaviour*, 33(1):21–44, 2003. DOI: 10.1111/1468-5914.00204. 34

E. Luger and T. Rodden. Terms of agreement: rethinking consent for pervasive computing. *Interacting with Computers*, 25(3):229–241, Feb. 2013a. ISSN 0953-5438. DOI: 10.1093/iwc/iws017. 76

E. Luger and T. Rodden. An informed view on consent for ubicomp. In *ACM international joint conference on Pervasive and ubiquitous computing (UbiComp '13)*, page 529, New York, USA, 2013b. ACM. ISBN 9781450317702. DOI: 10.1145/2493432.2493446. 76

P. Lukowicz, S. Pentland, and A. Ferscha. From context awareness to socially aware computing. *IEEE Pervasive Computing*, 11(1):32–41, 2012. ISSN 1536-1268. DOI: 10.1109/mprv.2011.82. 53, 54

D. Lyon. Terrorism and surveillance: security, freedom, and justice after September 11 2001. Privacy Lecture Series, November 12, 2001. URL `privacy.openflows.org/pdf/lyon_p aper.pdf`. 29

D. Lyon, editor. *Surveillance as social sorting: privacy, risk and automated discrimination*. Routledge, 2002. DOI: 10.4324/9780203994887. 29

M. Madden. Public Perceptions of Privacy and Security in the Post-Snowden Era, 2014. URL `http://www.pewinternet.org/2014/11/12/public-privacy-perceptions/`. 76, 79

B. Malin and L. Sweeney. How (not) to protect genomic data privacy in a distributed network: using trail re-identification to evaluate and design anonymity protection systems. *Journal of biomedical informatics*, 37(3):179–92, June 2004. ISSN 1532-0464. DOI: 10.1016/j.jbi.2004.04.005. 41

E. Malmi and I. Weber. You are what apps you use: demographic prediction based on user's apps. In *Tenth International AAAI Conference on Web and Social Media*, 2016. 76

E. Mandel. How the Napa earthquake affected bay area sleepers. The Jawbone Blog, Aug. 2014. URL `https://jawbone.com/blog/napa-earthquake-effect-on-sleep/`. 62

S. Mann. Wearable computing. In *The Encyclopedia of Human-Computer Interaction*, chapter 22. Interaction Design Foundation, 2nd ed. edition, 2013. `http://www.interaction-design.org/books/hci.html`. 46

S. Mann, J. Nolan, and B. Wellman. Sousveillance: inventing and using wearable computing devices for data collection in surveillance environments. *Surveillance & Society*, 1(3):331–355, July 2003. URL `http://www.surveillance-and-society.org/journalv1i3.htm`. DOI: 10.24908/ss.v1i3.3344. 30

A. Marthews and C. E. Tucker. Government surveillance and internet search behavior. Draft, SSRN, 2017. DOI: 10.2139/ssrn.2412564. 79

G. T. Marx. Murky conceptual waters: The public and the private. *Ethics and Information Technology*, 3(3):157–169, 2001. URL `web.mit.edu/gtmarx/www/murkypublicandprivate.html`. 37, 62

G. T. Marx. Some information age techno-fallacies. *Journal of Contingencies and Crisis Management*, 11(1):25–31, Mar. 2003. ISSN 0966-0879. DOI: 10.1111/1468-5973.1101005. 85

D. Mattioli. On Orbitz, Mac users steered to pricier hotels. *Wall Street Jounral*, `http://www.wsj.com/articles/SB10001424052702304458604577488822667325882`, August 2012. 67

V. Mayer-Schönberger. Generational development of data protection in Europe. In P. E. Agre and M. Rotenberg, editors, *Technology and Privacy: The New Landscape*, chapter 8, pages 219–242. The MIT Press, Cambridge, MA, USA, 1998. 18, 27

A. McClurg. A thousand words are worth a picture: a privacy tort response to consumer data profiling. *Nw. UL Rev.*, 98:63–1787, June 2003. ISSN 00293571. URL `http://scholar.google.com/scholar?hl=en{&}btnG=Search{&}q=intitle:A+Thousand+Words+Are+Worth+a+Picture:+A+Privacy+Tort+Response+to+Consumer+Data+Profiling{#}0`. 84, 85

A. M. McDonald and L. F. Cranor. The cost of reading privacy policies. *I/S: A Journal of Law and Policy for the Information Society*, 4(3):540–565, 2008. `https://kb.osu.edu/bitstream/handle/1811/72839/ISJLP_V4N3_543.pdf` 70, 73

R. McGarvey. Is your rental car company spying on you and your driving? Here's how they do it. *TheStreet.com*, mar 2015. URL `https://www.thestreet.com/story/13089306/1/is-your-rental-car-company-spying-on-you-and-your-driving-heres-how-they-do-it.html`. 60

C. R. McKenzie, M. J. Liersch, and S. R. Finkelstein. Recommendations implicit in policy defaults. *Psychological Science*, 17(5):414–420, 2006. DOI: 10.1037/e640112011-058. 74

R. Meyer. Facebook and the Cambridge analytica scandal, in 3 paragraphs. *The Atlantic*, Mar. 2018. URL `https://www.theatlantic.com/technology/archive/2018/03/the-camb ridge-analytica-scandal-in-three-paragraphs/556046/`. 29

Microsoft.   Cortana.   `http://www.microsoft.com/en-us/mobile/campaign-cortana/`, 2014. 48

G. R. Milne and M. J. Culnan. Strategies for reducing online privacy risks: why consumers read (or don't read) online privacy notices. *Journal of Interactive Marketing*, 18(3):15–29, Jan. 2004. ISSN 1094-9968. DOI: 10.1002/dir.20009. 70

S. Moncrieff, S. Venkatesh, and G. West. Dynamic privacy assessment in a smart house environment using multimodal sensing. *ACM Transactions on Multimedia Computing, Communications, and Applications*, 5(2), 2008. DOI: 10.1145/1413862.1413863. 50

Mozilla Foundation. Using geolocation, 2018. URL `https://developer.mozilla.org/en-US/docs/Web/API/Geolocation/Using_geolocation`. 63

E. Musk. A most peculiar test drive. Tesla blog, `http://www.teslamotors.com/blog/most-peculiar-test-drive`, 2013. 61

National Archives. America's founding documents. National Archives. URL `https://www.archives.gov/founding-docs`. 15

National Police Library – College of Computing. The effects of cctv on crime. Technical report, College of Policing, UK, 2013. URL `http://library.college.police.uk/docs/what-works/What-works-briefing-effects-of-CCTV-2013.pdf`. 21

S. Nisenbaum. Ways of being alone in the world. *The American Behavioral Scientist*, 27(6):785, 1984. DOI: 10.1177/000276484027006009. 34

H. Nissenbaum. Protecting privacy in an information age: the problem of privacy in public. *Law and Philosophy*, 17(5):559–596, 1998. DOI: 10.2307/3505189. 38

H. Nissenbaum. Privacy as contextual integrity. *Washington Law Review*, 79(1):119–159, 2004. URL `http://ssrn.com/abstract=534622`. 38, 72

H. Nissenbaum. *Privacy in Context - Technology, Policy, and the Integrity of Social Life*. Stanford University Press, 2009. ISBN 978-0804752367. DOI: 10.1080/15536548.2011.10855919. 35, 38, 39, 78, 79, 87

H. Nissenbaum. A contextual approach to privacy online. *Daedalus*, 140(4):32–48, Oct. 2011. ISSN 0011-5266. DOI: 10.1162/daed_a_00113. 78

P. A. Norberg, D. R. Horne, and D. A. Horne. The privacy paradox: personal information disclosure intentions versus behaviors. *Journal of Consumer Affairs*, 41(1):100–126, 2007. DOI: 10.1111/j.1745-6606.2006.00070.x. 22, 80

R. Nord, M. Barbacci, P. Clements, and R. Kazman. Integrating the architecture trade-off analysis method (ATAM) with the cost benefit analysis method (CBAM). Technical report, Software Engineering Institute (SEI), Pittsburgh, PA, USA, 2003. URL `http://repository.cmu.edu/sei/537/`. DOI: 10.21236/ada421615. 89

N. Notario, A. Crespo, Y.-S. Martin, J. M. D. Alamo, D. L. Metayer, T. Antignac, A. Kung, I. Kroener, and D. Wright. PRIPARE: integrating privacy best practices into a privacy engineering methodology. In *2015 IEEE Security and Privacy Workshops*, pages 151–158. IEEE, May 2015. ISBN 978-1-4799-9933-0. DOI: 10.1109/spw.2015.22. 89, 90

A. Nusca. How voice recognition will change the world. *SmartPlanet Online*, Nov. 2011. URL `https://www.zdnet.com/article/how-voice-recognition-will-change-the-world/`. 47

OECD. OECD guidelines on the protection of privacy and transborder flows of personal data. The Organisation for Economic Co-operation and Development, Sept. 1980. URL `http://www.oecd.org/sti/ieconomy/oecdguidelinesontheprotectionofprivacyandtransborderflowsofpersonaldata.htm`. 12, 85

OECD. The OECD privacy framework. The Organisation for Economic Co-operation and Development, 2013. URL `http://www.oecd.org/internet/ieconomy/privacy-guidelines.htm`. 12, 72, 81

M. C. Oetzel and S. Spiekermann. A systematic methodology for privacy impact assessments: a design science approach. *European Journal of Information Systems*, 23(2):126–150, Mar. 2014. ISSN 0960-085X. DOI: 10.1057/ejis.2013.18. 89

C. Palahniuk. *Fight Club*. W.W. Norton, New York, USA, 1996. ISBN 978-0393039764. 23

L. Palen and P. Dourish. Unpacking "privacy" for a networked world. In *Conference on Human factors in computing systems (CHI '03)*, pages 129–136, New York, USA, 2003. ACM. ISBN 1581136307. DOI: 10.1145/642611.642635. 79

S. Parakilas. We can't trust facebook to regulate itself. *The New York Times*, 2017. URL `https://www.nytimes.com/2017/11/19/opinion/facebook-regulation-incentive.html`. 22

R. Parasuraman, T. Sheridan, and C. Wickens. A model for types and levels of human interaction with automation. *IEEE Transactions on Systems, Man, and Cybernetics - Part A: Systems and Humans*, 30(3):286–297, May 2000. ISSN 10834427. DOI: 10.1109/3468.844354. 77, 81

B. Parducci, H. Lockhart, and E. Rissanen. eXtensible access control markup language (XACML) version 3.0. Committee Specification, OASIS, 2010. URL `http://docs.oasis-open.org/xacml/3.0/xacml-3.0-core-spec-cs-01-en.pdf`. DOI: 10.17487/rfc7061. 72

S. N. Patel, J. A. Kientz, G. R. Hayes, S. Bhat, and G. D. Abowd. Farther than you may think: an empirical investigation of the proximity of users to their mobile phones. In *UbiComp 2006: Ubiquitous Computing*, pages 123–140, 2006. DOI: 10.1007/11853565_8. 47

S. N. Patel, T. Robertson, J. A. Kientz, M. S. Reynolds, and G. D. Abowd. At the flick of a switch: detecting and classifying unique electrical events on the residential power line. In J. Krumm, G. D. Abowd, A. Seneviratne, and T. Strang, editors, *UbiComp 2007: Ubiquitous Computing*, pages 271–288, 2007. Springer Berlin Heidelberg. DOI: 10.1007/978-3-540-74853-3_16. 64

S. N. Patel, M. S. Reynolds, and G. D. Abowd. Detecting human movement by differential air pressure sensing in HVAC system ductwork: an exploration in infrastructure mediated sensing. In J. Indulska, D. Patterson, T. Rodden, and M. Ott, editors, *Pervasive Computing. Pervasive 2008*, pages 1–18, 2008. Springer. DOI: 10.1007/978-3-540-79576-6_1. 64

P. Pelegris, K. Banitsas, T. Orbach, and K. Marias. A novel method to detect heart beat rate using a mobile phone. In *Engineering in Medicine and Biology Society (EMBC), 2010 Annual International Conference of the IEEE*, pages 5488–5491, Aug. 2010. DOI: 10.1109/iembs.2010.5626580. 47

J. W. Penney. Chilling effects: Online surveillance and wikipedia use. *Berkeley Technology Law Journal*, 31(1):117, 2016. URL `https://ssrn.com/abstract=2769645`. 79

Pew Research Center. An analysis of android app permissions. `http://www.pewinternet.org/2015/11/10/an-analysis-of-android-app-permissions/`, 2015. 76

B. Pfleging, S. Schneegass, and A. Schmidt. Multimodal interaction in the car. In A. L. Kun, L. Boyle, B. Reimer, A. Riener, J. Healey, W. Zhang, B. Pfleging, and M. Kurz, editors, *Proceedings of the 4th International Conference on Automotive User Interfaces and Interactive Vehicular Applications - AutomotiveUI '12*, pages 155–162, New York, USA, 2012. ACM Press. ISBN 9781450317511. DOI: 10.1145/2390256.2390282. 50

R. W. Picard. Affective computing: challenges. *International Journal of Human-Computer Studies*, 59(1-2):55–64, July 2003. ISSN 10715819. DOI: 10.1016/s1071-5819(03)00052-1. 53

J. P. Pickett, editor. *The American Heritage College Dictionary*. Houghton Mifflin Co, 4th edition, Apr. 2002. 25

S. Platt, Ed. *Respectfully Quoted: A Dictionary of Quotations Requested from the Congressional Research Service*. Library of Congress, Washington, DC, 1989. http://www.bartleby.com /73/ 98

B. Popper. Google announces over 2 billion monthly active devices on Android. *The Verge Website (verge.com)*, may 2017. URL https://www.theverge.com/2017/5/17/15654454/androi d-reaches-2-billion-monthly-active-users. 22

R. C. Post. Three concepts of privacy. *Georgetown Law Journal*, 89:2087, 2001. URL http://digitalcommons.law.yale.edu/cgi/viewcontent.cgi?article=1184& context=fss_papers. DOI: 10.1016/j.clsr.2015.05.010. 7

President's Concil of Advisors on Science and Technology. Big data and privacy: a technological perspective. Report to the President, Executive Office of the President, May 2014. URL https://obamawhitehouse.archives.gov/sites/default/files/micro sites/ostp/PCAST/pcast_big_data_and_privacy_-_may_2014.pdf 76

Privacy Rights Clearinghouse. A review of the fair information principles: the foundation of privacy public policy, Feb. 2004. URL http://www.privacyrights.org/ar/fairinfo.ht m. 11

Privacy Rights Clearinghouse. California medical privacy fact sheet C4: your prescriptions and your privacy. https://www.privacyrights.org/fs/fsC4/CA-medical-prescription- privacy, July 2012. 66

W. Prosser. Privacy. *California Law Journal*, 48:383–423, 1960. As cited in Solove and Rotenberg [2003]. DOI: 10.2307/3478805. 17

A. Rao, F. Schaub, and N. Sadeh. What do they know about me? Contents and concerns of online behavioral profiles. In *PASSAT '14: Sixth ASE International Conference on Privacy, Security, Risk and Trust*, 2014. 66, 67

R. Rehmann. Der weg aus der hooligan-datenbank. Tages-Anzeiger, Nov. 2014. URL http://www.tagesanzeiger.ch/schweiz/standard/Der-Weg-aus-der- HooliganDatenbank/story/21957458. 2

J. R. Reidenberg, T. Breaux, L. F. Cranor, B. French, A. Grannis, J. T. Graves, F. Liu, A. M. McDonald, T. B. Norton, R. Ramanath, N. C. Russell, N. Sadeh, and F. Schaub. Disagreeable privacy policies: mismatches between meaning and users' understanding. *Berkeley Technology Law Journal*, 30, 2015. URL http://papers.ssrn.com/abstract=2418297. DOI: 10.2139/ssrn.2418297. 75

J. R. Reidenberg, J. Bhatia, T. D. Breaux, and T. B. Norton. Ambiguity in privacy policies and the impact of regulation. *The Journal of Legal Studies*, 45(S2):S163–S190, 2016. URL https://doi.org/10.1086/688669. 75

H. T. Reis, P. Shaver, et al. Intimacy as an interpersonal process. *Handbook of personal relationships*, 24(3):367–389, 1988. 34

M. Reissenberger. 50 jahre bundesverfassungsgericht: Volkszählung. DeutschlandRadio Schwerpunktthema, January 4, 2004. URL `http://www.dradio.de/homepage/schwerpunkt-verfassungsgericht-010904.html`. 27

C. Roda. *Human attention in digital environments*. Cambridge University Press, 2011. ISBN 9780521765657. URL `http://www.cambridge.org/ch/academic/subjects/psychology/cognition/human-attention-digital-environments?format=HB{&}isbn=9780521765657`. DOI: 10.1017/cbo9780511974519. 51

F. Roesner, D. Molnar, A. Moshchuk, T. Kohno, and H. J. Wang. World-driven access control for continuous sensing. In *Proceedings of the 2014 ACM SIGSAC Conference on Computer and Communications Security*, CCS '14, pages 1169–1181, New York, USA, 2014. ACM. ISBN 978-1-4503-2957-6. DOI: 10.1145/2660267.2660319. 80

N. A. Romero, P. Markopoulos, and S. Greenberg. Grounding privacy in mediated communication. *Computer Supported Cooperative Work (CSCW)*, 22(1):1–32, 2013. ISSN 0925-9724. DOI: 10.1007/s10606-012-9177-z. 79

J. Rosen. A watchful state. *The New York Times Magazine*, Oct. 2001. `http://www.nytimes.com/2001/10/07/magazine/a-watchful-state.html`. 21

I. S. Rubinstein. Big data: The end of privacy or a new beginning? *International Data Privacy law*, 3(2):74–87, 1 May 2013. DOI: 10.1093/idpl/ips036. 85

S. Ruggieri, D. Pedreschi, and F. Turini. Integrating induction and deduction for finding evidence of discrimination. *Artificial Intelligence and Law*, 18(1):1–43, Mar. 2010. ISSN 0924-8463. DOI: 10.1007/s10506-010-9089-5. 85

B. Rössler. *Der Wert des Privaten*. Suhrkamp Verlag, Frankfurt/Main, Germany, 2001. 34

N. Sadeh, J. Hong, L. F. Cranor, I. Fette, P. Kelley, M. Prabaker, and J. Rao. Understanding and capturing people's privacy policies in a mobile social networking application. *Personal and Ubiquitous Computing*, 13(6):401–412, Aug. 2009. ISSN 1617-4909. DOI: 10.1007/s00779-008-0214-3. 77, 80

F. Sadri. Ambient intelligence: a survey. *ACM Computing Surveys*, 43(4):1–66, Oct. 2011. ISSN 03600300. DOI: 10.1145/1978802.1978815. 50

R. Saleh, D. Jutla, and P. Bodorik. Management of users' privacy preferences in context. In *International Conference on Information Reuse and Integration*, pages 91–97. IEEE, Aug. 2007. ISBN 1-4244-1499-7. DOI: 10.1109/iri.2007.4296603. 80

P. Samuelson.   Privacy as intellectual property?   *Stanford Law Review*, 52:1125–1173, 2000.   URL `http://www.sims.berkeley.edu/~pam/papers/privasip_draft.pdf`. DOI: 10.2307/1229511. 25

M. Satyanarayanan. Fundamental challenges in mobile computing. In *Proceedings of the fifteenth annual ACM symposium on Principles of distributed computing*, pages 1–7. ACM, 1996. URL `http://dl.acm.org/citation.cfm?id=248053`.   00813. DOI: 10.1145/248052.248053. 45, 46, 47

M. Satyanarayanan. Pervasive computing: vision and challenges. *IEEE Personal Communications*, 8(4):10–17, 2001. ISSN 10709916. DOI: 10.1109/98.943998. 51, 52

F. Schaub. *Dynamic Privacy Adaptation in Ubiquitous Computing*. Doctoral dissertation, University of Ulm, Ulm, Germany, 2014. 76, 77

F. Schaub.   Context-adaptive privacy mechanisms.   In A. Gkoulalas-Divanis and C. Bettini, editors, *Handbook of Mobile Data Privacy*, chapter 13, pages 337–372. Springer, Dec. 2018. ISBN 978-3-319-98160-4. DOI: 10.1007/978-3-319-98161-1_13. 79

F. Schaub, R. Deyhle, and M. Weber. Password entry usability and shoulder surfing susceptibility on different smartphone platforms. In *11th International Conference on Mobile and Ubiquitous Multimedia (MUM '12)*, page 1, New York, USA, 2012. ACM. ISBN 9781450318150. DOI: 10.1145/2406367.2406384. 41

F. Schaub, B. Könings, P. Lang, B. Wiedersheim, C. Winkler, and M. Weber. PriCal: context-adaptive privacy in ambient calendar displays. In *UbiComp '14 Proceedings of the 2014 ACM International Joint Conference on Pervasive and Ubiquitous Computing*, pages 499–510. ACM Press, 2014. ISBN 9781450329682. DOI: 10.1145/2632048.2632087. 77, 80

F. Schaub, B. Könings, and M. Weber. Context-adaptive privacy: leveraging context awareness to support privacy decision making. *IEEE Pervasive Computing*, 14(1):34–43, 2015. DOI: 10.1109/mprv.2015.5. 74, 76, 77, 79, 81

F. Schaub, R. Balebako, and L. F. Cranor.   Designing effective privacy notices and controls.   *IEEE Internet Computing*, 21(3):70–77, May–June 2017.   ISSN 1089-7801. DOI: 10.1109/mic.2017.265102930. 70, 74, 75, 76

B. W. Schermer. The limits of privacy in automated profiling and data mining. *Computer Law & Security Review*, 27(1):45–52, Feb. 2011. ISSN 0267-3649. DOI: 10.1016/j.clsr.2010.11.009. 83, 85

B. Schilit, N. Adams, and R. Want.   Context-aware computing applications.   In *Workshop on Mobile Computing Systems and Applications*, pages 85–90. IEEE, 1994. ISBN 0-8186-6345-6. DOI: 10.1109/mcsa.1994.512740. 48

J. Schiller. *Mobile communications*. Pearson Education, 2003. 46

A. Schmidt. Implicit human computer interaction through context. *Personal Technologies*, 4 (2-3):191–199, June 2000. ISSN 0949-2054. DOI: 10.1007/bf01324126. 52

A. Schmidt. Context-aware computing: context-awareness, context-aware user interfaces, and implicit interaction. In M. Soegaard and R. F. Dam, editors, *Encyclopedia of Human-Computer Interaction*, chapter 14, pages 1–28. The Interaction-Design.org Foundation, Aarhus, Denmark, 2012. URL http://www.interaction-design.org/encyclopedia/context-aw are_computing.html. 48

A. Schmidt, M. Beigl, and H.-W. Gellersen. There is more to context than location. *Computers & Graphics*, 23(6):893–901, Dec. 1999. ISSN 00978493. DOI: 10.1016/s0097-8493(99)00120-x. 48

A. Schmidt, B. Pfleging, F. Alt, A. Sahami, and G. Fitzpatrick. Interacting with 21st-century computers. *IEEE Pervasive Computing*, 11(1):22–31, Jan. 2012. ISSN 1536-1268. DOI: 10.1109/mprv.2011.81. 45

B. Schneier. Security risks of frequent-shopper cards. Schneier on Security, Feb. 18 2005. URL https://www.schneier.com/blog/archives/2005/02/security_risks.html. 1

V. Schouberechts. *The Post Book – History of the European post in 50 exclusive documents*. Lanoo Publishers, jun 2016. 32

B. Schwartz. The social psychology of privacy. *American Journal of Sociology*, 73(6):741–752, May 1968. ISSN 0002-9602. DOI: 10.1086/224567. 39

P. M. Schwartz. Property, privacy, and personal data. *Harvard Law Review*, 117(7):2055–2128, 2004. URL http://ssrn.com/abstract=721642. DOI: 10.2307/4093335. 66

C. Scully, J. Lee, J. Meyer, A. Gorbach, D. Granquist-Fraser, Y. Mendelson, and K. Chon. Physiological parameter monitoring from optical recordings with a mobile phone. *Biomedical Engineering, IEEE Transactions on*, 59(2):303–306, Feb. 2012. ISSN 0018-9294. DOI: 10.1109/tbme.2011.2163157. 47

C. Seife. 23andMe is terrifying; but not for the reasons the FDA thinks. *Scientific American*, Sept. 2013. https://www.scientificamerican.com/article/23andme-is-terr ifying-but-not-for-the-reasons-the-fda-thinks/. 33

P. Sieghart. *Privacy and Computers*. Latimer, London, 1976. 58

G. Silberg and R. Wallace. Self-driving cars: the next revolution. Technical report, KPMG LLP and the Center for Automotive Research (CAR), 2012. URL https://home.kpmg.com/be /en/home/insights/2012/08/self-driving-cars-the-next-revolution.html. 50

S. Silverstein. What price loyalty? Los Angeles Times, Feb. 1999. URL `http://articles.l atimes.com/1999/feb/07/business/fi-5738/`. DOI: 10.1093/jiplp/jpp199. 1

A. Simmel. Privacy. In *International Encyclopedia of the Social Sciences*, page 480. MacMillan, 1968. As cited in Cate [1997]. 36

H. A. Simon. *Models of bounded rationality: Empirically grounded economic reason*, MIT press, 1982. 73

H. J. Smith, T. Dinev, and H. Xu. Information privacy research: an interdisciplinary review. *MIS Quarterly*, 35(4):989–1015, 2011. DOI: 10.2307/41409970. 73

D. J. Solove. Privacy self-management and the consent dilemma. *Harvard Law Review*, 126: 1880–1903, 2013. URL `http://ssrn.com/abstract=2171018`. 22, 36, 65, 70, 76

D. J. Solove. A taxonomy for privacy. *University of Pennsylvania Law Review*, 154(3):477–560, 2006. URL `http://ssrn.com/abstract=667622`. DOI: 10.2307/40041279. 40, 43

D. J. Solove. *Understanding Privacy*. Harvard University Press, 2008. ISBN 978-0674027725. 39, 40, 42, 43, 65

D. J. Solove and W. Hartzog. The FTC and the new common law of privacy. *Columbia Law Review*, 114(583), 2014. URL `https://ssrn.com/abstract=2312913`. DOI: 10.2139/ssrn.2312913. 17

D. J. Solove and M. Rotenberg. *Information Privacy Law*. Aspen Publishers, New York, USA, 1st edition, 2003. 10, 13, 14, 15, 28, 100, 117

D. J. Solove and P. M. Schwartz. An overview of privacy law. In *Privacy Law Fundamentals*, chapter 2. IAPP, 2015. URL `https://ssrn.com/abstract=2669879`. 15

D. J. Solove and P. M. Schwartz. *Information Privacy Law*. Wolters Kluwer, 6th edition, 2018. ISBN 978-1454892755. 43

S. Spiekermann and L. F. Cranor. Engineering privacy. *IEEE Transactions on Software Engineering*, 35(1):67–82, Jan. 2009. ISSN 0098-5589. DOI: 10.1109/tse.2008.88. 86

S. Spiekermann, J. Grossklags, and B. Berendt. E-privacy in 2nd generation e-commerce: privacy preferences versus actual behavior. In *Proceedings of the 3rd ACM conference on Electronic Commerce*, pages 38–47. ACM, 2001. DOI: 10.1145/501158.501163. 22

P. Sprenger. Sun on privacy: 'Get over it.' *Wired News*, Jan. 1999. URL `https://www.wired. com/1999/01/sun-on-privacy-get-over-it/`. 24

M. Staples, K. Daniel, M. Cima, and R. Langer. Application of micro- and nano-electromechanical devices to drug delivery. *Pharmaceutical Research*, 23(5):847–863, 2006. ISSN 0724-8741. DOI: 10.1007/s11095-006-9906-4. 50

T. Starner. Project Glass: An extension of the self. *IEEE Pervasive Computing*, 12(2):14–16, Apr. 2013. ISSN 1536-1268. DOI: 10.1109/mprv.2013.35. 50

Statista. Number of apps available in leading app stores as of 3rd quarter 2018. `http://www.statista.com/statistics/276623/number-of-apps-available-in-leading-app-stores/`, October 2018. 48

B. Stone. Amazon erases Orwell books from Kindle devices, July 2009. URL `https://www.nytimes.com/2009/07/18/technology/companies/18amazon.html`. 60

W. J. Stuntz. The substantive origins of criminal procedure. *The Yale Law Journal*, 105(2):393, Nov. 1995. ISSN 00440094. DOI: 10.2307/797125. 25

L. Sweeney. k-anonymity: a model for protecting privacy. *International Journal of Uncertainty, Fuzziness and Knowledge-Based Systems*, 10(5):557–570, 2002. ISSN 0218-4885. DOI: 10.1142/s0218488502001648. 41

P. Swire and K. Ahmad. *Foundations of Information Privacy and Data Protection: A Survey of Global Concepts, Laws and Practices*. The International Association of Privacy Professionals (IAPP), 2012. ISBN 978-0-9795901-7-7. 20

O. Tene and J. Polonetsky. Privacy in the age of big data - a time for big decisions. *Stanford Law Review Online*, 64(63), 2012. URL `https://www.stanfordlawreview.org/online/privacy-paradox-privacy-and-big-data/`. 85

O. Tene and J. Polonetsky. Big data for all: Privacy and user control in the age of analytics. *Northwestern Journal of Technology and Intellectual Property*, 11(5):239, sep 2013. URL `https://papers.ssrn.com/sol3/papers.cfm?abstract_id=2149364`. 85

Th. Sc. Community. *Pervasive Adaptation: The Next Generation Pervasive Computing Research Agenda*. Institute for Pervasive Computing, Johannes Kepler University Linz, Linz, 2011. ISBN 9783200022706. URL `http://perada.eu/essence`. 54

The Economist. When smart becomes spooky. *The Economist Online*, Nov. 2014. URL `https://www.economist.com/news/2014/11/13/when-smart-becomes-spooky`. 61

The Economist. The signal and the noise. Technical report, London, UK, Mar. 2016. URL `https://www.economist.com/sites/default/files/20160326_tech_politics.pdf`. 29

The State of New York. Article 240 - NY penal law, 2018a. URL `http://ypdcrime.com/penal.law/article240.htm`. 23

The State of New York. Article 490 - NY penal law, 2018b. URL `http://ypdcrime.com/pen al.law/article490.htm`. 23

E. Toch.   Super-Ego: a framework for privacy-sensitive bounded context-awareness.   In *5th ACM International Workshop on Context-Awareness for Self-Managing Systems (CASE-MANS '11)*, pages 24–32, New York, USA, 2011. ACM.   ISBN 9781450308779. DOI: 10.1145/2036146.2036151. 77

E. Toch. Crowdsourcing privacy preferences in context-aware applications. *Personal and Ubiquitous Computing*, 18(1):129–141, Jan. 2014. ISSN 1617-4917. DOI: 10.1007/s00779-012-0632-0. 77

L. Tsai, P. Wijesekera, J. Reardon, I. Reyes, S. Egelman, D. Wagner, N. Good, and J.-W. Chen. Turtle Guard: helping Android users apply contextual privacy preferences. In *Thirteenth Symposium on Usable Privacy and Security (SOUPS 2017)*, pages 145–162, Santa Clara, CA, 2017. USENIX Association. ISBN 978-1-931971-39-3. URL `https://www.usenix.org/confe rence/soups2017/technical-sessions/presentation/tsai`. 77, 81

J. Turow, M. Hennessy, and N. Draper. The tradeoff fallacy: How marketers are misrepresenting american consumers and opening them up to exploitation. Technical report, Annenberg School for Communication, University of Pennsylvania, 2015. DOI: 10.2139/ssrn.2820060. 76

C. Ulbrich. Can spam? Or new can of worms? *Wired News*, Dec. 2003. URL `https://www. wired.com/2003/12/can-spam-or-new-can-of-worms/`. 26

United Nations. Universal declaration of human rights. Adopted and proclaimed by General Assembly resolution 217 A (III) of December 10, 1948. URL `http://www.un.org/Overv iew/rights.html`. 10

United States Government. E-government act of 2002. Public Law 107-347, Dec. 2002. URL `https://www.gpo.gov/fdsys/pkg/PLAW-107publ347/content-detail.html`. 89

B. Ur, J. Jung, and S. Schechter. Intruders versus intrusiveness: teens' and parents' perspectives on home-entryway surveillance. In *UbiComp 2014*, pages 129–139. ACM Press, 2014. ISBN 9781450329682. DOI: 10.1145/2632048.2632107. 61

A. Valdez. Everything you need to know about Facebook and Cambridge analytica. Wired, March 23, 2018, 2018. URL `https://www.wired.com/story/wired-facebook-cambri dge-analytica-coverage/`. 79

W. G. Voss. European Union data privacy law reform: general data protection regulation, privacy shield, and the right to delisting. *Business Lawyer,*, 72(1):221–233, Jan. 2017. URL `https://ssrn.com/abstract=2894571`. 43

J. Waldo, H. S. Lin, and L. I. Millett. Engaging privacy and information technology in a digital age. Technical report, National Research Council, 2007. URL `http://www.nap.edu/open book.php?record_id=11896`. DOI: 10.29012/jpc.v2i1.580. 62

Y. Wang and M. Kosinski. Deep neural networks are more accurate than humans at detecting sexual orientation from facial images. *Journal of Personality and Social Psychology*, 114(2):246–257, 2018. DOI: 10.31234/osf.io/hv28a. 84

R. Want. Introduction to ubiquitous computing. In J. Krumm, editor, *Ubiquitous Computing Fundamentals*, chapter 1. CRC Press, Boca Raton, FL, 2010. DOI: 10.1201/9781420093612.ch1. 45, 49

S. D. Warren and L. D. Brandeis. The right to privacy. *Harvard Law Review*, 4(5):193–220, Dec. 1890. ISSN 0017811X. DOI: 10.2307/1321160. 8, 9, 33

Washington State Legislature. Electronic communication devices, chapter 19.300 RCW, 2009. URL `http://app.leg.wa.gov/RCW/default.aspx?cite=19.300`. 16

M. Weiser. The computer for the 21st century. *Scientific American*, 265(3):94–104, Jan. 1991. ISSN 1536-1268. DOI: 10.1038/scientificamerican0991-94. 47, 49, 50, 51, 52, 57, 58

M. Weiser. Some computer science issues in ubiquitous computing. *Communications of the ACM*, 36(7):75–84, 1993. DOI: 10.1145/159544.159617. 47, 49

M. Weiser and J. Brown. The coming age of calm technology. In P. J. Denning and R. M. Metcalfe, editors, *Beyond Calculation: The Next Fifty Years of Computing*, volume 8. Springer, 1997. DOI: 10.1007/978-1-4612-0685-9_6. 47, 51, 53

M. A. Weiss and K. Archick. U.S. - EU Data privacy: from Safe Harbor to Privacy Shield. Technical report, Congressional Research Service, Washington, D.C., USA, May 2016. URL `https://epic.org/crs/R44257.pdf`. 19

R. Wenning, M. Schunter, L. F. Cranor, B. Dobbs, S. Egelman, G. Hogben, J. Humphrey, M. Langheinrich, M. Marchiori, M. Presler-Marshall, J. Reagle, and D. A. Stampley. The platform for privacy preferences 1.1 (P3P1.1) specification. W3C working group note, W3C, 2006. URL `http://www.w3.org/TR/P3P11/`. 72, 80

A. F. Westin. *Privacy and Freedom*. Atheneum, New York, 1967. DOI: 10.2307/1339271. 11, 26, 32, 33, 34, 35, 36, 37, 43, 73

P. Wijesekera, A. Baokar, L. Tsai, J. Reardon, S. Egelman, D. Wagner, and K. Beznosov. The feasibility of dynamically granted permissions: aligning mobile privacy with user preferences. In *2017 IEEE Symposium on Security and Privacy (SP)*, pages 1077–1093, May 2017. DOI: 10.1109/sp.2017.51. 80, 81

P. Wijesekera, J. Reardon, I. Reyes, L. Tsai, J.-W. Chen, N. Good, D. Wagner, K. Beznosov, and S. Egelman. Contextualizing privacy decisions for better prediction (and protection). In *Proceedings of the 2018 CHI Conference on Human Factors in Computing Systems*, CHI '18, pages 268:1–268:13, New York, USA, 2018. ACM. ISBN 978-1-4503-5620-6. DOI: 10.1145/3173574.3173842. 77

Wikipedia. Jeremy Bentham (Aug. 21, 2004, 17:37 UTC). In *Wikipedia: The Free Encyclopedia*. Wikimedia Foundation, 2004. URL en.wikipedia.org/wiki/Jeremy_Bentham. 14

L. E. Willis. Why not privacy by default. *Berkeley Technology Law Journal*, 29(1), 2014. URL http://scholarship.law.berkeley.edu/cgi/viewcontent.cgi?article=2019{&}context=btlj. DOI: 10.2139/ssrn.2349766. 71

Wired News. Due process vanishes in thin air. *Wired News*, April 8, 2003. URL http://www.wired.com/news/print/0,1294,58386,00.html. 30

D. Wright. Should privacy impact assessments be mandatory? *Communications of the ACM*, 54 (8):121–131, Aug. 2011. ISSN 00010782. DOI: 10.1145/1978542.1978568. 88

D. Wright. Making privacy impact assessment more effective. *The Information Society*, 29(5):307–315, 2013. URL https://doi.org/10.1080/01972243.2013.825687. DOI: 10.1080/01972243.2013.825687. 88

D. Wright and P. De Hert, editors. *Privacy Impact Assessment*. Springer Netherlands, Dordrecht, 2012. ISBN 978-94-007-5402-7. DOI: 10.1007/978-94-007-2543-0. 39

G. Yee. A privacy-preserving ubicomp architecture. In *International Conference on Privacy, Security and Trust (PST '06)*, New York, USA, 2006. ACM. ISBN 1595936041. DOI: 10.1145/1501434.1501467. 80

Authors' Biographies

MARC LANGHEINRICH

Marc Langheinrich is full professor in the Faculty of Informatics at the Università della Svizzera Italiana (USI) in Lugano, Switzerland. His research focuses on privacy in mobile and pervasive computing systems, in particular with a view towards social compatibility. Other research interests include usable security and pervasive displays. Marc is a member of the Steering Committee of the UbiComp conference series and chairs the IoT conference Steering Committee. He has been a General Chair or Program Chair of most major conferences in the field—including Ubicomp, PerCom, Pervasive, and the IoT conference—and currently serves as the Editor-in-Chief for *IEEE Pervasive Magazine*. Marc holds a Ph.D. from ETH Zürich, Switzerland. He can be reached at `langheinrich@acm.org`. For more information, see `https://uc.inf.usi.ch/`.

FLORIAN SCHAUB

Florian Schaub is an assistant professor in the School of Information and the Computer Science and Engineering Division at the University of Michigan. His research focuses on understanding and supporting people's privacy and security behavior and decision making in complex socio-technological systems. His research interests span privacy, human-computer interaction, and emergent technologies, such as the Internet of Things. Florian received a doctoral degree in Computer Science from the University of Ulm, Germany, and was a postdoctoral fellow in Carnegie Mellon University's School of Computer Science. His research has been honored with Best Paper Awards at CHI, the ACM SIGCHI Conference on Human Factors in Computing, and SOUPS, the Symposium on Usable Privacy and Security. Florian can be reached at `fschaub@umich.edu`. For more information, see `https://si.umich.edu/people/florian-schaub`.

Printed in the United States
by Baker & Taylor Publisher Services